MW00837963

ELECTROMAGNETIC ANECHOIC CHAMBERS

IEEE Press
445 Hoes Lane, P.O. Box 1331
Piscataway, NJ 08855-1331

ELECTROMAGNETIC ANECHOIC CHAMBERS

A Fundamental Design and Specification Guide

LELAND H. HEMMING

IEEE Electromagnetic Compatibility Society, *Sponsor*
IEEE Antennas and Propagation Society, *Sponsor*

IEEE Press

A JOHN WILEY & SONS, INC., PUBLICATION

This text is printed on acid-free paper. ♾

Published simultaneously in Canada.

For ordering and customer service, call 1-800-CALL-WILEY

Library of Congress Cataloging in Publication Data is available.

ISBN 0-471-20810-8

10 9 8 7 6 5 4 3 2 1

CONTENTS

FOREWORD

At last, a handbook for electromagnetic anechoic chambers! A single source for fundamentals, design, specification, and testing of anechoic chambers written by one of the leading experts in anechoic chamber design and practice. Leland Hemming has brought together many disparate references to fill a void in the art of electromagnetic measurement: the anechoic chamber.

This handbook presents all major types of anechoic chambers including: rectangular, tapered, and double horn. It includes all major measurement techniques such as far-field, compact, and near-field testing. It also addresses all major types of measurements, including: antenna, radar cross section, electromagnetic compatibility, radiated susceptibility, and radiated emissions.

A key component of electromagnetic anechoic chambers is the material used to cover all interior surface of the chamber. The design of a chamber hinges on the proper selection and placement of the absorber on the various surfaces of the chamber, including the back wall, the side-wall specular zones, the floor, and corners of the chamber. The handbook includes in-depth discussion of the types of absorber, the shapes of absorber, their reflectivity properties, and the measurement techniques used to test the absorber.

Overlooked issues such as lighting, ventilation, fire protection, high power testing, and electromagnetic shielding are important in the final design and specification of an anechoic chamber. This handbook supplies this information as well.

This handbook also includes the important topics of chamber test procedures and acceptance testing. The testing techniques presented include: absorber testing, free space VSWR, pattern comparison, X-Y scanner, and RCS evaluation. The basis for each technique and the procedure to carry out each technique is presented. I have found that without periodic testing and documentation, and subsequent refurbishment, chamber performance can degrade significantly.

I have become acutely aware of the scarcity of information regarding the design, proper use, performance measurement and specification for anechoic chambers, in my teaching and research on antenna, radome and radar cross section measurement techniques over the last 35 years. This is especially true for the newer measurement techniques such as compact and near-field measurements performed in an anechoic chamber. It is with great pleasure that there is now an authoritatively written handbook that can supply the needed information for this important part of electromagnetic measurement.

Hemming has chosen to include only the well-established, time-tested techniques and information. I recommend this handbook to all students and practitioners of electromagnetic measurement.

EDWARD B. JOY

Professor Emeritus
Georgia Institute of Technology

PREFACE

I first became involved in the design and/or procurement of anechoic chambers in 1972 when I was responsible for procuring a dual-mode anechoic test facility. The facility was designed to house a compact range for apertures up to 1.5 m and a low-frequency tapered chamber that could operate down to 200 MHz. The physical arrangement that we developed utilized a common antenna test positioner located in the test region of the tapered chamber. The back wall of the tapered chamber was designed to open up and reveal the compact range reflector. Since that time, I have collected material on anechoic materials and chambers. Subsequently, I became chief designer for two different anechoic chamber manufacturers over a 10-year period. In 1985, I returned to my primary career activity as an antenna engineer. However, I have kept in touch with the anechoic chamber field by working as a part-time consultant and have kept my files current on the technology. This book is based on that experience.

The need for indoor testing of electromagnetic radiating devices, which began in the early 1950s [see Chapter 1, Ref. 1], has led to a number of companies providing chambers and absorber products supporting a range of electromagnetic testing requirements. The testing requirements range from hand-held telephone antennas to whole vehicle testing of automobiles or aircraft. In recent years, the proliferation of portable electronic devices has led to the potential for extensive electromagnetic interference between various products. Because of this potential interference, various government agencies throughout the world have had to set up emission and immunity test requirements. This has led to the extensive use of indoor electromagnetic test facilities [see Chapter 2, Ref. 2]. These facilities, including the more conventional chambers used for testing of intentional radiators, such as antennas, continue to add to the number of anechoic test chambers worldwide. In the aerospace industry, it is common to test missiles, aircraft, or similar weapons platforms for their radar cross-section in an attempt to reduce detectability by means of radar. All of these various testing requirements have led to a series of specialized test facilities to accomplish the required testing. This book reviews the current state of the art in indoor electromagnetic testing facilities and their design and specification.

A glossary relating to electromagnetic measurements and anechoic chambers is included at the back of the book, so that the reader will be able to follow the terminology used in conjunction with a particular anechoic chamber design.

The purpose of this handbook is to provide the designer/procurer of electromagnetic chambers with a single source of practical information on the full range of anechoic chamber designs. Included are chapters on a large variety of anechoic chambers used for a broad range of electromagnetic measurements that are commonly conducted in indoor test facilities. Sufficient information is given on measurement theory to support the chamber design procedures provided in each of the specific chamber designs. Test facilities for the measurement of antennas, scattering (RCS,) and electromagnetic compatibility are detailed, for a variety of anechoic chamber configurations. An extensive set of photographs is provided and demonstrates the broad range of anechoic test facilities that have been built to measure various types of equipment. A special color section is provided which highlights some of the more interesting anechoic test facilities that have been built to solve various measurement problems. Finally, design/procurement checklists are provided for each of the various chamber configurations.

I am solely responsible for the technical information included in this handbook. This information was solicited from the many suppliers who responded to my request for detailed data on their absorber products and for pictures of their anechoic chamber installations.

ACKNOWLEDGMENTS

I thank my wife, Valda, for her unending support during the three years it took to complete this project. I am especially thankful for the help provided by Donald J. Martin, a co-worker at Boeing-Mesa, who processed all the artwork and photographs into digital format and created many of the drawings. Many thanks go to Edward Pelton, my former supervisor at McDonnell Douglas Technologies, who proofread the final manuscript, and for his suggestions on some of the technical and presentational issues. Finally, my thanks to the volunteer reviewers for their suggestions and recommendations on the various revisions of the book, to the IEEE Press, and to the John Wiley & Sons staff who worked with me on the final preparation and publication of this design/procurement guide.

LELAND H. HEMMING

Mesa, Arizona
May 2002

ELECTROMAGNETIC ANECHOIC CHAMBERS

CHAPTER 1

INTRODUCTION

1.1 THE TEXT ORGANIZATION

Chapter 2 is devoted to the principles of electromagnetic measurements that pertain to anechoic chamber design. The concept of plane waves, uniform fields, and uniform phase are given to establish the basis for determining the properties of a radiating or scattering device in free space. Also discussed is the impact of the measurement site itself on the accuracy of the measurements. For a later comparison, the principles involved in the design of outdoor test ranges are described. References specific to each chamber concept are included in the individual sections. General references applicable to the general field of electromagnetic test facilities are included in the Selected Bibliography.

Next, in Chapter 3, radio-frequency and microwave absorbing materials are described with detailed information on their performance and how they are used to establish controlled electromagnetic test environments within an indoor test facility.

Chapter 4 provides information on the chamber enclosure. If electromagnetic shielding is required, common available systems are described with references detailing their construction.

Chapters 5 through 8 detail the designs of the various test facilities broken down by the geometry of the test facility. These include the rectangular chamber, the compact range chamber, and those chambers where geometry is important in the design, including the tapered chamber, the double horn chamber, and the TEM cell.

Chapter 9 summarizes all the test procedures associated with electromagnetic acceptance testing of the various chambers and also the testing of absorber materials used in the construction of the chambers.

Chapter 10 provides a summary of the types of indoor test facilities that have been developed for the various electromagnetic measurements. These include chambers for testing antennas, radar cross section, electromagnetic compatibility, and electromagnetic systems. Extensive use of photographs has been used to illustrate these various anechoic chambers. A special insert of full-color photographs has been included to highlight the design concepts used in anechoic chamber design.

Appendix C provides a series of design and specification checklists for the various types of anechoic chambers.

A selected bibliography is given in support of the content of the book.

An extensive index is provided so that convenient reference can be made to any subject covered by the text.

REFERENCES

1. W. H. Emerson, Electromagnetic Wave Absorbers and Anechoic Chambers Through the Years, *IEEE Transactions on Antennas and Propagation,* Vol. AP-21, No. 4, July 1973.
2. B. F. Lawrence, Anechoic Chambers, Past and Present, *Conformity,* Vol. 6, No.4, pp. 54–56, April 2000.
3. IEEE Std 145-1983, *IEEE Standard Definitions of Terms for Antennas,* IEEE Press, New York, 1983.
4. ANSI C63.14: 1998, American National Standard Dictionary for Technologies of Electromagnetic Compatibility (EMC), Electromagnetic Pulse (EMP), and Electrostatic Discharge (ESD).
5. IEEE 100, *The Authoritative Dictionary of IEEE Standard Terms,* 7th edition, IEEE Press, New York, 2000.

CHAPTER 2

MEASUREMENT PRINCIPLES PERTAINING TO ANECHOIC CHAMBER DESIGN

2.1 INTRODUCTION

The purpose of the book is to provide a guide on the design and performance specification of electromagnetic anechoic chambers or indoor test facilities. However, some knowledge of electromagnetic measurements is required to ensure that all applicable design factors have been applied in any given design situation.

The measurement of electromagnetic waves involves a large number of electronic devices. These devices can be categorized as intentional or unintentional radiators. Measurement of intentional radiators (such as antennas) or the scattering of electromagnetic energy, as in radar cross section, requires specialized testing facilities. Determining the level of radiation from unintentional radiators such as digital devices or determining the level of immunity an electronic device has with respect to an impinging electromagnetic wave also involves the measurement of electromagnetic waves in testing facilities designed specifically for the measurement to be performed.

Electromagnetic waves result from the acceleration of electric charges. The electric field due to an unaccelerated charge (one at rest or in uniform motion in a straight line) is radially directed and decreases as the square of the distance from the charge. However, the acceleration of the charge gives rise to a tangential component of the electric field, and this decreases linearly with distance [1]. This time-varying electric field has associated with it a time varying magnetic field; together, they comprise an electromagnetic field. An electromagnetic field that decreases linearly with distance represents an outward radiation.

In practice, one is almost always concerned with macroscopic effects resulting from acceleration of gross numbers of charges. On the macroscopic scale, the interrelationship between electric and magnetic fields is described mathematically by Maxwell's equations [2, 3]. An additional set of equations called constitutive relationships [3, 4] specifies the characteristics of the medium in which the field exists.

The mathematics of electromagnetic fields and the associated media makes use of vector and tensor analysis. The analysis can become very involved in many problems, especially those involving propagation in nonisotropic, nonreciprocal, nonlinear, or nonhomogeneous media. Although certain problems in electromagnetic measurements can require application of more detailed mathematics, only the relatively elementary aspects of vector analysis and electromagnetic theory are used in this book. The reader is assumed to be familiar with basic vector operations and concepts of wave phenomena. Standard texts, such as the references cited in this and subsequent chapters, should be consulted as needed.

2.2 MEASUREMENT OF ELECTROMAGNETIC FIELDS

2.2.1 Introduction

The variety of electromagnetic field measurement techniques is fairly extensive. The following examples represent a sampling of the most common types of measurement and their parameters, which are affected by the testing environment.

2.2.2 Antennas

2.2.2.1 Introduction. Antenna measurements are extensively conducted throughout industry and government test facilities. These measurements involve essentially two basic parameters. The first is the distribution of the radiated energy in space about the antenna or antenna pattern and its associated gain. The second parameter is how well the antenna is matched to the transmission line feeding the antenna or input impedance.

2.2.2.2 Antenna Patterns. The energy radiated from an antenna is a three dimensional problem. The pattern shape can be nearly spherical or all the energy can be directed in one direction, such as that generated by a large reflector antenna. The facilities for measuring antennas vary greatly, depending on the purpose of the antenna system and the type of pattern it generates. Low-gain or electrically small antennas can be measured in simple rectangular chambers in the microwave frequency range, but require specially designed chambers below 1 GHz, due to the absorption properties of the anechoic material that must be used at the lower frequencies. High-gain antennas require very long length outdoor ranges or the use of compact antenna ranges indoors. Each of these different facilities approximates the ideal test environment to varying degrees of performance. The differences will be discussed in the following chapters in conjunction with the design of the particular facility. The following is general information that applies to all facilities.

Figure 2.1 depicts a generalized antenna pattern. A major lobe is shown with its half-power beamwidth noted. The sidelobe levels are also noted. The gain of the antenna is referenced to the peak of the pattern. The accuracy of the measured pattern is a function of the level of extraneous energy existing in the test region of the antenna test facility. If the extraneous level is –45 dB, then the 20-dB sidelobe will have a ±1-dB uncertainty associated with the measurement. Figure 2.2 is used to evaluate the degree of uncertainty that applies to a given measurement, knowing the level of extraneous energy that exists in the test region and a given pattern level.

2.2.2.3 Antenna Gain. The IEEE Test Procedure for Antennas, Std-149 [5], defines the *power gain* of an antenna in a specified direction as 4π times the ratio

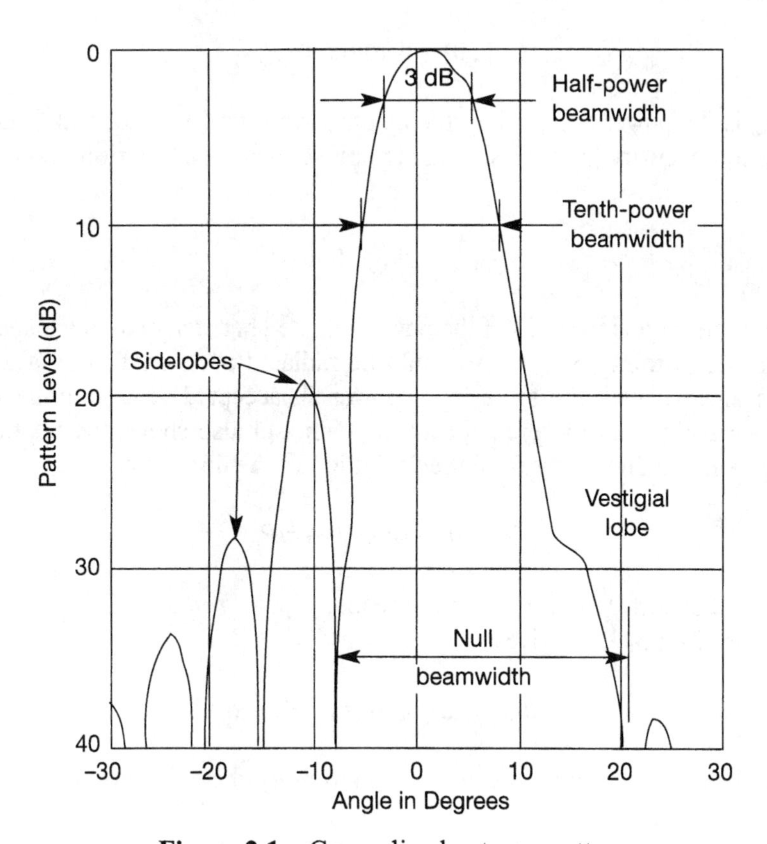

Figure 2.1 Generalized antenna pattern.

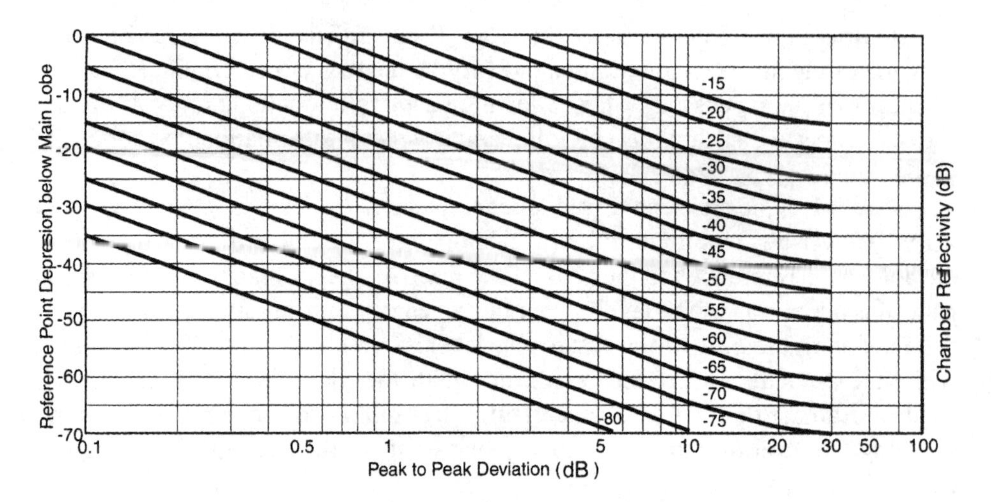

Figure 2.2 Pattern level uncertainty versus extraneous signal level.

of the power radiated per unit solid angle in that direction to the net power accepted by the antenna from its generator. This is described mathematically by

$$G(\phi, \theta) = 4\pi\Phi(\phi, \theta)/P_o \tag{2-1}$$

where P_o is the power accepted by the antenna from its generator and $\Phi(\phi, \theta)$ is the radiation intensity (power radiated per unit solid angle). Rewriting (2-1) in the form

$$G(\phi, \theta) = \Phi(\phi, \theta)4\pi/P_o \tag{2-2}$$

shows that the gain is the ratio of the power radiated per steradian in the specified direction (ϕ, θ) to the power that would be radiated per steradian by a lossless isotropic antenna with the same input power P_o accepted at its terminals. This form gives a physical interpretation to the gain and also shows the relationship between gain and directivity from the definition of the directivity

$$G(\phi, \theta) = \Phi(\phi, \theta)4\pi/P_r \tag{2-3}$$

where P_r is the power *radiated* by the antenna.

Dividing (2-2) by (2-3) gives

$$G(\phi, \theta)/D(\phi, \theta) = P_r/P_o = \eta \tag{2-4}$$

Because the efficiency can never be as great as unity, the gain must always be less than the directivity.

Techniques for measuring gain can be found in Ref. 5.

2.2.2.4 Antenna Impedance. The input impedance of an antenna is influenced by the physical environment adjacent to the antenna. Antennas radiate energy; and if the energy is reflected back into the antenna, then the apparent input impedance changes. If the reflecting object is greater than approximately one wavelength away from the antenna, then the measurements are fairly representative of the actual input impedance. If the measurement is to be conducted within a small RF absorber lined enclosure, then the reflectivity of the absorber must be selected carefully as pointed out in Chapter 3.

2.2.3 Radiated Emissions

The measurement of emissions from electronic devices has become a very important part of bringing an electronic device to market. Throughout the world, government agencies now specify the maximum emissions that an electronic device can radiate. The levels are very low due to the presence of so many devices in a given office environment. Careful calibration of the measurement site is essential because large errors could lead to expensive equipment fixes if the specified levels of emissions are not met. Currently, these measurements are conducted on open area test sites (OATS) or in alternative indoor sites such as anechoic chambers. These measurements involve the measurement of the site attenuation—that is, the propagation loss between the device under test and the measurement antenna [6]. Knowing this loss and using a spectrum analyzer, the level of the emission at a given distance can be determined. These measurements are generally conducted at range lengths varying from 3 to 30 m. Anechoic chambers are commonly used for the 3-m range length. Larger companies have built a number of 10-m chambers [7].

The measurement aspect that is most important in EMI anechoic design is the degree to which the site attenuation is the same as measured on an OATS. Because the facility is confined within an enclosure, the repeatability of the measurement within the test volume must also be verified. A third item common to all EMI measurements is the uncertainty associated with the antennas used in the measurements and how they are calibrated.

2.2.4 Radiated Susceptibility

Radiated susceptibility requires the measurement to be conducted in a shielded anechoic chamber because high levels of power are required to generate the fields necessary to test the susceptibility of a device to various levels of electromagnetic fields. Careful design of these facilities is required to minimize uncertainty in the measurements. Detailed information is given on the design of these specialized test facilities. The uniformity of the field illuminating the device under test (DUT) is the most critical parameter in susceptibility measurements [6].

2.2.5 Military Electromagnetic Compatibility

Military radiated electromagnetic measurements are conducted in shielded anechoic enclosures using test procedures and absorber treatments specified in MIL-STD-461E [8].

2.2.6 Antenna System Isolation

Antenna system isolation measurements require a very low extraneous signal environment. A large clear area around the vehicle under test, if located outdoors, is required. A large test volume is required in an anechoic chamber. Indoors, the site isolation is the sum of the two-way path loss to the walls plus the reflectivity of the absorber. The sum should be at least 6 dB greater than the system requirements to provide an adequate signal-to-noise margin for the measurement. A special case of this type of measurement is the measurement of passive intermodulation (PIM) [9].

2.2.7 Radar Cross Section

The measurement of radar cross section has special test environment requirements. These include (1) minimizing of forward scatter from the transmitter to the test region, (2) minimizing extraneous energy reflected from surrounding surfaces, (3) uniform illumination of the target, in phase and amplitude, and (4) the ability to operate over very broad frequency ranges [10]. Each of these items is discussed in the sections describing a particular test chamber concept.

2.3 FREE-SPACE TEST REQUIREMENTS

2.3.1 Introduction

The radiation characteristics of an electronic device in spherical coordinates under a given set of conditions can be described by the functions

$$\begin{aligned} &G1(\phi, \theta), G2(\phi, \theta) \\ &\delta(\phi, \theta) \\ &\eta \\ &Z \end{aligned} \tag{2-5}$$

where $G1$ and $G2$ are the gain functions for two orthogonal polarizations, δ is the phase angle between the output signals for the two polarizations, η is the efficiency of the device determined by its ohmic losses, and Z is the impedance that the device presents at its input terminals. These relationships are illustrated in Figure 2.3.

Only in the case of an intentional radiator, such as an antenna, are these parameters easily determined. In addition, environmental factors must be considered in any electromagnetic measurement problem, and suitable allowance must be made for their effects.

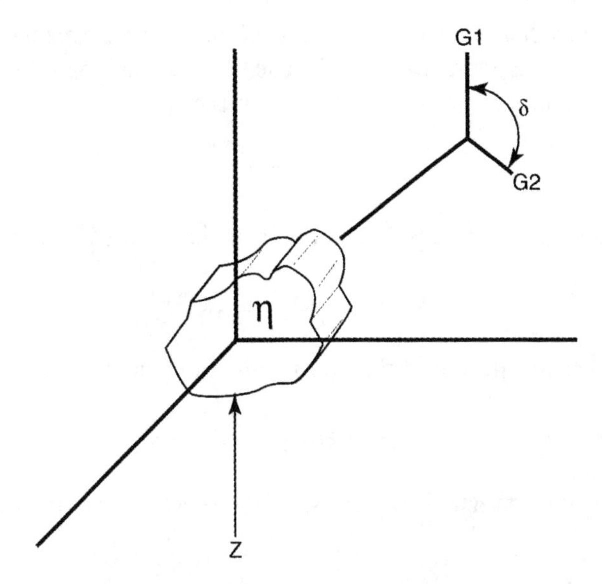

Figure 2.3 Symbolic relationship of measured parameters.

Before proceeding to the specifics of electromagnetic testing that is performed in anechoic chambers, it is appropriate to review certain fundamental terms and relations that are basic to electromagnetic measurements.

2.3.2 Phase

A commonly employed criterion for determining the minimum allowable separation between the source antenna and the antenna under test is to restrict the phase variation to be less than a maximum of $\pi/8$ radian, or 22.5 degrees [11].

Consider a source of spherical waves and an antenna under test (AUT) located at a distance R away from it as illustrated in Figure 2.4. The largest phase differ-

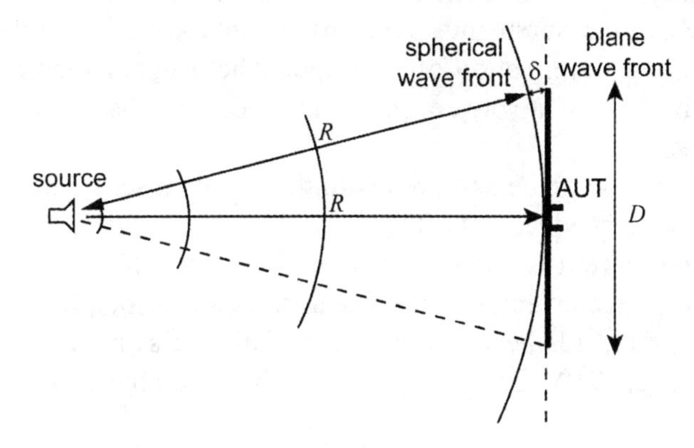

Figure 2.4 Derivation of the far-field criteria.

ence between the spherical wave and the ideal plane wave appears at the edges of the AUT, which corresponds to the difference in the wave path lengths, δ. Under the common assumption, $\kappa\delta$ fulfills the requirement

$$\kappa\delta \leq \pi/8 \tag{2-6}$$

The difference in the wave path lengths, δ, is determined by noticing that

$$(R + \delta)^2 = R^2 + (D_{max}/2)^2 \tag{2-7}$$

The only physical solution of the above quadratic equation for δ is

$$\delta = (R^2 + (D_{max}/2)^2)^{1/2} - R \tag{2-8}$$

The above is approximated by the use of the binomial expansion (the first two terms only) as

$$\delta = R[(1 + (D_{max}/2R)^2 - 1] = R[1 + 1/2(D_{max}/2R)^2 - 1] = D^2_{max}/4R \tag{2-9}$$

The minimum distance from the source of the spherical wave is now determined from the requirement in (2-6),

$$\kappa D^2_{max}/4R = (2\pi/\lambda)(D^2_{max}/4R) \leq \pi/8 \tag{2-10}$$

Under this condition,

$$R_{min} = 2D^2_{max}/\lambda \tag{2-11}$$

If antenna measurements are made at this range length, then there will be a minor departure of the nulls of the radiation pattern and the location and levels of the minor lobes from their infinite-range values.

Equation (2-11) was derived as a guide to setting up an antenna measurement range to accurately measure antenna radiation patterns. When conducting radar cross-section (RCS) measurements, concerns are not the pattern of the source and receive antenna, but illumination of the target. The primary requirement is that the source antenna has relatively constant amplitude and phase characteristics, as seen at the target.

These requirements can be achieved with the target closer to the antennas than the value calculated by equation (2-11).

No published criterion is known for determining exactly how close the target should be to the source antenna. Some guidance was provided by Hu [12], where the phase within the 3-dB beamwidth is essentially the same as at infinity for a distance of $R/4$ or $0.5D^2/\lambda$. Equation (2-11) can be generalized to

$$R = KD^2/\lambda \tag{2-12}$$

When a target accuracy of 1 dB is required, Knott [13] has shown that K should be between 2 and 10, depending on the target characteristics. The types of targets that require a large value of K are those in which the edge scattering is comparable to the specular-scattering contribution. A convenient chart for determining the target size as a function of K is given in Figure 2.5.

Radiated emission and susceptibility measurements are usually conducted for both the near and far fields due to the broad frequency coverage and the distances set by the testing requirements [14, 15].

2.3.3 Amplitude

For accuracy in simulated far zone measurements, the illuminating field must be sufficiently constant in amplitude both along the line of sight and in planes normal to the line of sight. If the field uniformity is not sufficient along the axis, then measurement errors will occur, especially in the minor lobes. If the amplitude taper across the aperture is too great, then the measured patterns will vary significantly from free-space. For high-performance antenna ranges the taper is generally held to be less than 0.25 dB, resulting in a measured gain error of less than 0.1 dB [11]. For a range length of $R = KD^2/\lambda$ [see equation (2-12)], the source diameter would be restricted to

$$d \leq 0.37\, KD \tag{2-13}$$

where d is the diameter of the source antenna, K is a constant, and D is the diameter of the AUT.

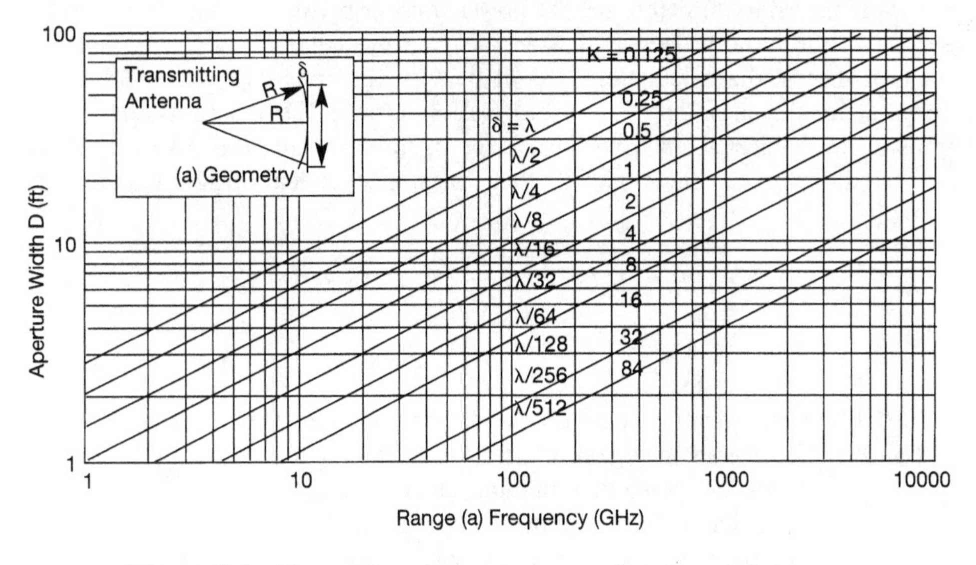

Figure 2.5 Test region size versus range length and frequency.

For most antenna measurements, if the amplitude variation is held to less than 1 dB, then the error is usually acceptable.

2.3.4 Polarization

At a large distance from a radiating antenna, the electric (E) and magnetic (H) vectors of the radiated field are at right angles to each other and to the direction of propagation. The two fields oscillate in time phase, and the ratio of their magnitudes (E/H) is a constant ζ, the intrinsic impedance of free space, and has a value of approximately 120π ohms.

The polarization of an electromagnetic field is described in terms of the direction in space of the electric field [11]. If the vector, which describes the electric field at a point in space, is always directed along a line, which is necessarily normal to the direction of propagation, the field is said to be *linearly* polarized. In general, however, the terminus of the electric vector describes an ellipse, and the field is said to be *elliptically* polarized.

Radiating devices exhibit polarization properties in relation to the fields they radiate or receive. If an antenna is operated in the receiving mode, it will not in general be polarization-matched to the incident field. If it is polarization-matched, it will extract maximum power from the field, and its polarization efficiency is said to be unity. If its polarization is orthogonal to the field, it will extract zero power, and its polarization efficiency is consequently zero.

The polarization properties of fields and antennas are of primary focus in all problems concerned with communication between antennas.

2.3.5 The Friis Transmission Formula

If it were necessary to determine the power transfer between two antennas by resorting to the basic processes that are defined by the field equations and diffraction theory, the calculations would be virtually impossible for all but the simplest of antennas. The Friis transmission formula [16] permits the power transfer to be determined from knowledge of the measured directive properties and the dissipative attenuation of the antennas, independent of their detailed design.

$$P_r \equiv P_o G(\phi, \theta) G(\phi', \theta')(\lambda/4\pi R)^2 \Gamma \tag{2-14}$$

P_r = power received,
P_o = power transmitted,
$G(\phi, \theta)$ = transmit antenna gain in direction of other antenna,
$G'(\phi', \theta')$ = receive antenna gain in direction of other antenna,
λ = operating wavelength in dimensions of R,
R = range length in meters or feet or inches,
Γ = polarization efficiency, a measure of how well the antennas, are polarization-matched.

2.4 SUPPORTING MEASUREMENT CONCEPTS

2.4.1 Introduction

When conducting electromagnetic measurements it is often required to establish a coordinate system, because the measurements are generally conducted in three dimensions. Also, most electromagnetic measurements are conducted in decibels (dB), because the dynamic range of the measurements often exceed a power ratio of more than million times or 60 dB. Another problem is that the measurements are conducted near the earth or in an enclosure, thus the problem of reflected energy must be addressed. These three general concepts are discussed in the following sections.

2.4.2 Coordinate Systems and Device Positioners

Almost all electromagnetic radiation measurement involves determination, in one way or another, of signal levels as a function of position or direction in space, usually the latter. Because of the nature of radiation, the spherical coordinate system (Figure 2.6) is the system that is most often employed in electromagnetic measurement problems. Often, the measurement of direction must be made to a

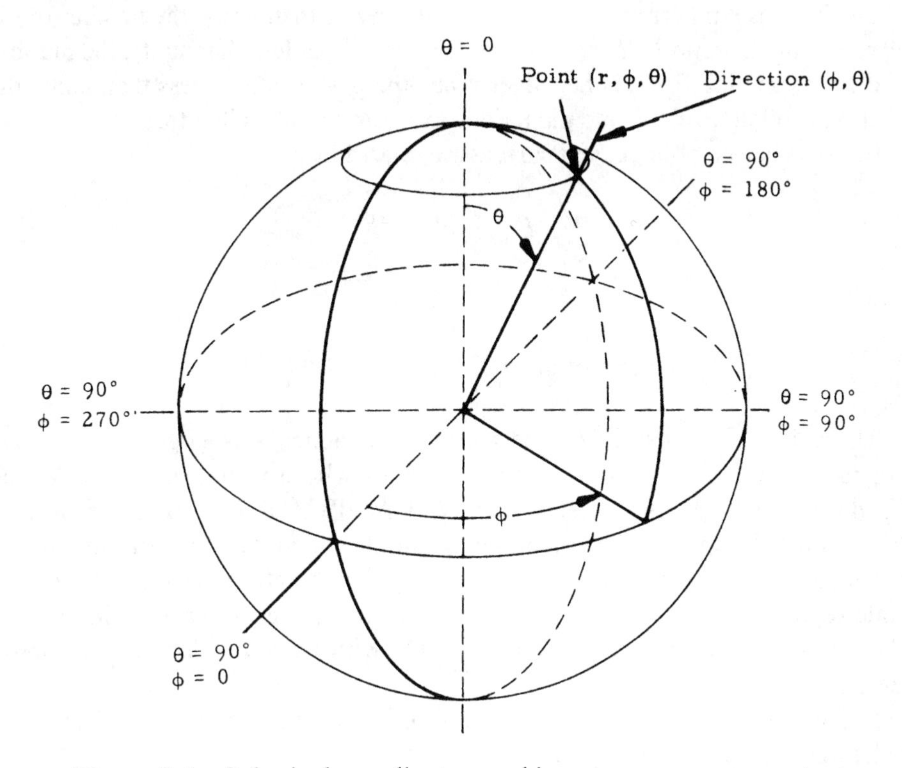

Figure 2.6 Spherical coordinates used in antenna measurements.

high degree of precision because of the requirements of the operational system of which the radiating device is a part.

Because of the relatively large distances that are often required between the device under test and the source antenna or sampling antenna (depending on whether the device is tested on receiving or transmitting), it is often not practical to explore the radiation pattern of the device in (ϕ, θ) by movement of the device or the antenna over a spherical surface. Instead, the line of sight between the test device and the measurement antenna is held fixed in space, while the device under test is changed in orientation to simulate movement of the line of sight over the sphere. This requirement has led to the development of special-purpose positioners, which are described for each test situation.

2.4.3 Decibels

The use of decibels in calculations and measurements of radiation measurements is almost essential, due to the large dynamic range of the quantities described. The decibel, abbreviated dB, is a logarithmic unit and is used to measure the ratio between two amounts of power. By definition,

$$\text{Number of dB} = 10 \log P_1/P_2 \tag{2-15}$$

where P_1/P_2 is a power ratio. When P_1/P_2 is greater than unity, the number of dB representing the ratio P_1/P_2 is positive. When P_1/P_2 is less than unity, the number of dB representing P_1/P_2 is negative. When the power ratio is less than unity, the fraction is often inverted, and the ratio is expressed as a decibel loss.

Because power, voltage, and current are related by

$$P = V^2/R = I^2R \tag{2-16}$$

we obtain

$$\text{Number of dB} = 20 \log V_1/V_2 = 20 \log I_1/I_2$$

The value of the use of decibels in electromagnetics is largely based upon two factors. First, if n_1 and n_2 are power ratios whose values in dB are N_1 and N_2, the product n_1n_2 is represented by $N_1 + N_2$ dB, and n_1/n_2 is represented by $N_1 - N_2$ dB. This permits the handling of products and quotients of large power ratios simply by the operations of addition and subtraction. Second, the decibel scale represents a compression of the power ratio scale, which permits the display of large power ratios on a single graph, with equal resolution at all power levels.

It is convenient to convert the Friis transmission formula (2-14) to decibels in order to simplify its application:

$$L_r = L_o + g(\phi, \theta) + g'(\phi, \theta) - 20 \log(4\pi R/\lambda) \tag{2-17}$$

where
L_r is the signal level at the output terminals of the receiving antenna in dBm,
L_o is the signal level at the input terminals of the transmitting antenna in dBm,
g is the gain of the source antenna,
g' is the gain of the receive antenna,
$g(\phi, \theta) = 10 \log G(\phi, \theta)$,
$g'(\phi', \theta') = 10 \log G'(\phi, \theta)$, and
R is the transmitter–receiver separation and λ = wavelength.
Note: R and λ must be in the same units (usually meters) of measure.
dBm is the signal level in power referenced to 1 mW.

2.4.4 Effects of Reflected Energy

When two or more coherent electromagnetic signals combine in the test region of an electromagnetic test facility, the fields illuminating the device under test are distorted and the accuracy of the measurement is degraded.

The discussion here assumes a separation, R, between sources and receiving antenna equal to or greater than $2D^2/\lambda$ (see Section 2.3.2), where D is the maximum dimension of the receiving aperture. It is further assumed that the ratio D/R is small in comparison with the half-power beamwidth of the source antenna's pattern, so that plane wave approximations are meaningful.

The mechanism by which a reflected signal interferes with the desired direct-path signal and distorts the incident field is demonstrated as follows. Consider the case of a direct-path plane wave of amplitude E_D that is normally incident on a test aperture, as shown in Figure 2.7. Let a reflected plane wave of amplitude E_R enter the aperture at an angle θ from the normal. The radial line defined by the unit vector **u** is the intersection of the plane containing the directions of propagation of the direct-path and reflected signals and the plane containing the aperture. The line having unit vector **v** is at some general angle α from **u** in the plane of the test aperture. At any given time t, the phase of the direct wave is constant across the aperture, and the magnitude of the direct-path field may be expressed in phasor notation as

$$E_D = E_D e^{j(\phi + \omega t)} \tag{2-18}$$

The phase of the postulated reflected plane wave will vary across the aperture so that the magnitude of the reflected field along the line **v** or any line parallel to **v** is given by

$$E_R = E_R e^{j(\phi + \alpha\chi + 2\pi/\lambda l \sin\theta \cos\alpha)} \tag{2-19}$$

In (2-18) and (2-19), φ and φ' are constants, λ is the wavelength, and l is a distance measured along the radial line **v** or along some other line parallel to **v**. The

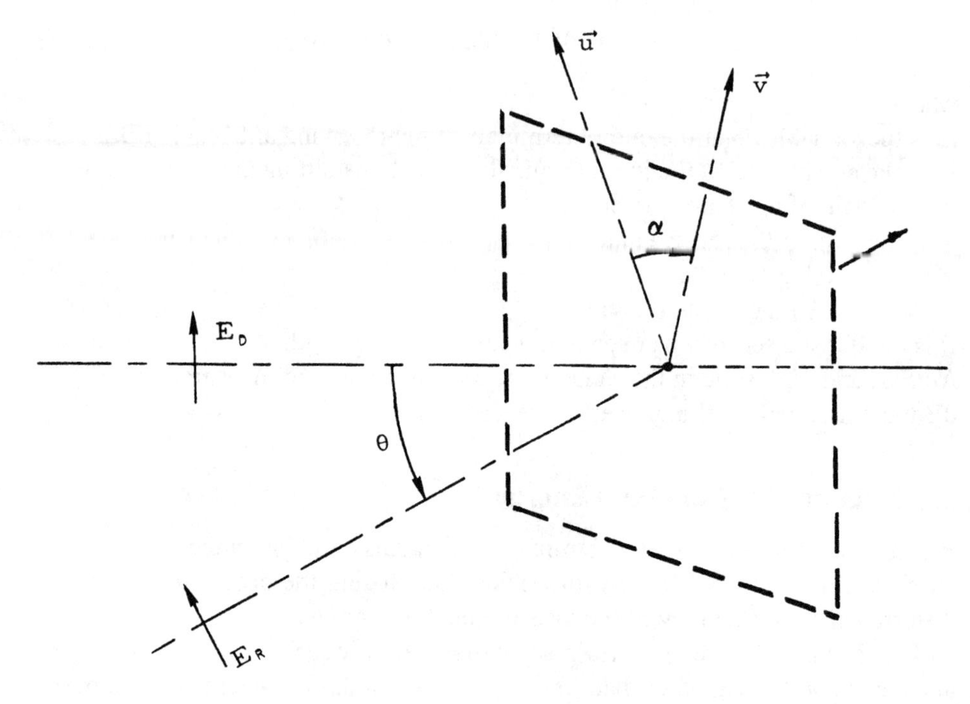

Figure 2.7 Summation of two coherent signals.

magnitude of the total field measured as a function of distance along **v** will be proportional to

$$E(\nu) = E_D + E_R \sin(2\pi/\lambda l \sin\theta \cos\alpha) \tag{2-20}$$

The distance between successive peaks of the resultant sinusoidal field is

$$P = \lambda/(\sin\theta \cos\alpha) \tag{2-21}$$

This spatial period or pitch has a minimum value of

$$P_{\min} = \lambda/(\sin\theta) \tag{2-22}$$

along the direction **u**, as shown in Figure 2.8.

While the spatial periods of the measured field fluctuations are functions of the position of the radial line in the aperture for which a recording of the field is made using a field probe, the peak-to-peak amplitude of these fluctuations is a constant proportional to

$$(E_D + E_R) - (E_D - E_R) = 2E_R \tag{2-23}$$

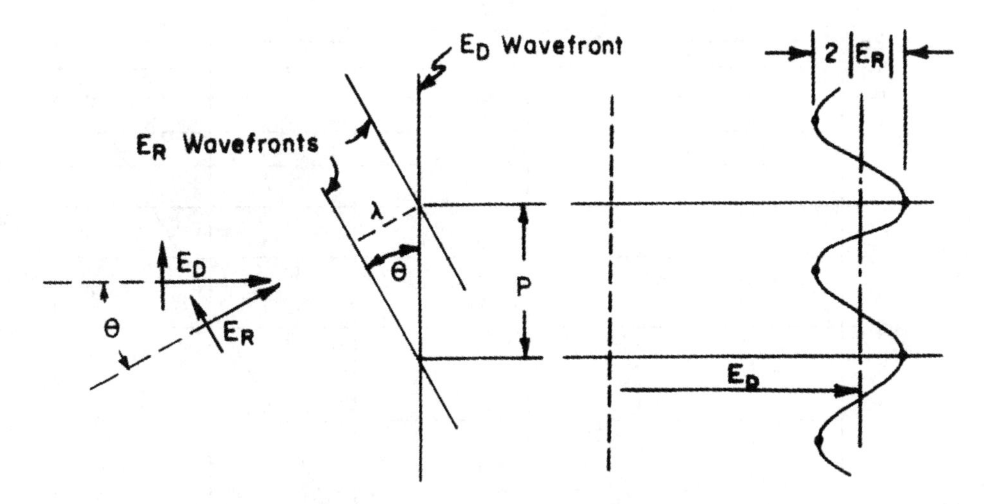

Figure 2.8 Vector addition of two coherent signals.

For typical logarithmic patterns, the ratio of E_R and E_D would be

$$E_R/E_D(\mathrm{dB}) = 20\log[(-1 + 10^{(\sigma/20)})/(1 + 10^{(\sigma/20)})] \qquad (2\text{-}24)$$

where σ is the difference in decibels between maxima and minima of the measured pattern. A plot of E_R/E_D versus σ is given in Figure 2.9.

The preceding example demonstrates the manner in which reflected signals would distort an otherwise planar wave front. In a more realistic case, neither the direct wave nor the reflected wave would be strictly planar, and there would be many sources of extraneous signals that could contribute to the total aperture field. However, the simple model presented above is quite useful in interpreting field probe data for the typical case where one reflected signal is somewhat stronger than other signals that may be present. An alternative method of performing the same analysis is given in Chapter 9.

2.4.5 Effects of Antenna Coupling

At the lower frequencies, the effects of inductive coupling between the source and the test antenna must be considered [11]. The ratio of the amplitude of the induction field to that of the radiation field is given as

$$\rho_\varepsilon = \lambda/(2\pi R) \qquad (2\text{-}25)$$

If one assumes $R > 10\lambda$, then $20\log(\rho_\varepsilon)$ is < -36 dB.

Mutual coupling due to scattering and reradiation of the energy by the test and source antennas is also of concern. If coupling does occur, frequency pulling and

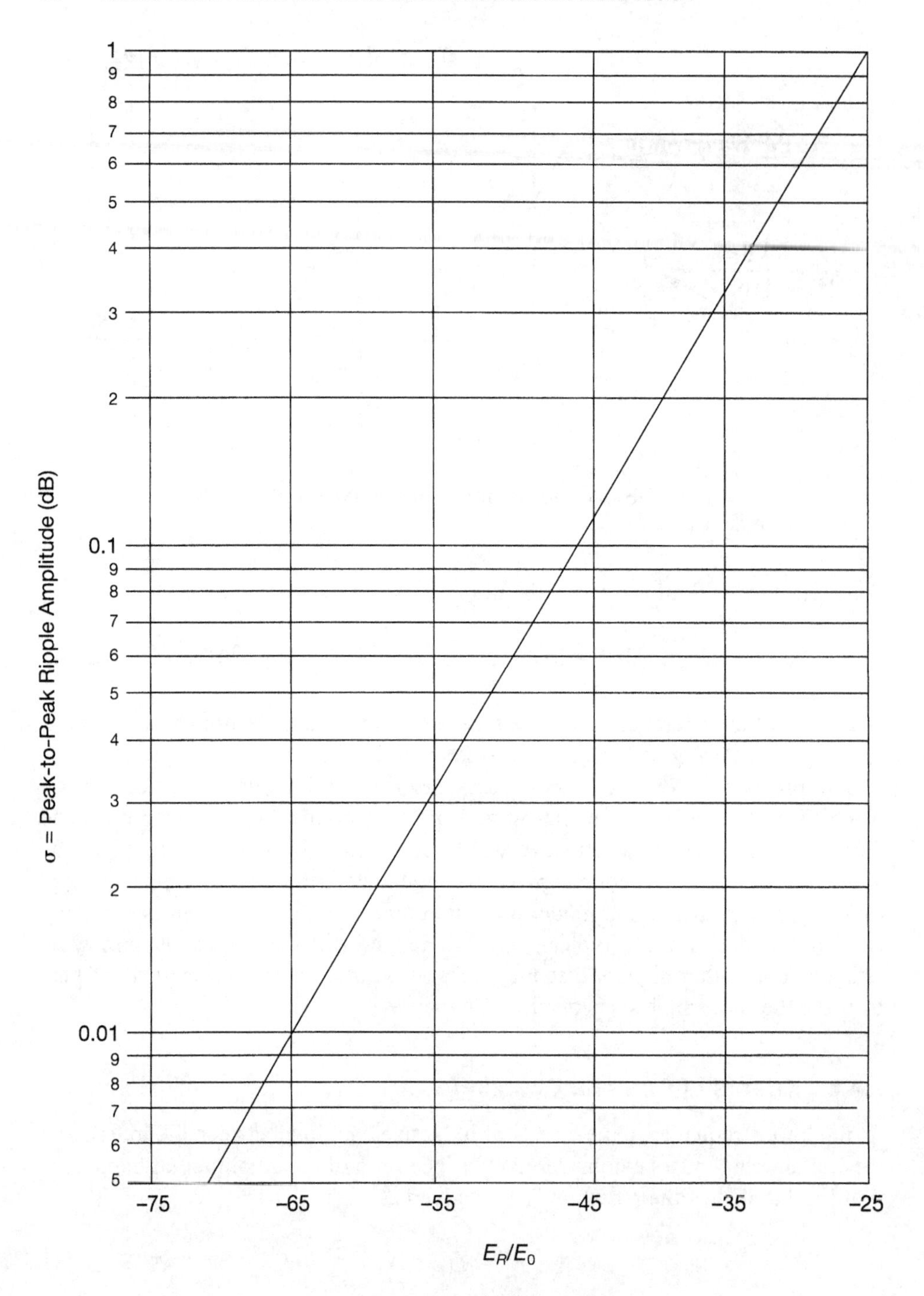

Figure 2.9 Plot of error versus extraneous signal level.

power level changes can be reduced by inserting an isolator or attenuator between the source and its antenna.

2.5 OUTDOOR MEASUREMENT FACILITIES

2.5.1 Introduction

Before proceeding into the design of anechoic chambers, a brief review of outdoor test ranges is required to set the stage.

The ideal test environment for determining far-zone performance would provide for a plane wave of uniform amplitude to illuminate the test aperture. Various approaches to simulation of this ideal electromagnetic environment have led to the evolution of two basic types of antenna test ranges: (1) free-space ranges and (2) reflection ranges.

Free-space ranges are those in which an attempt is made to suppress or remove the effects of all surroundings, including the range surface or surfaces along the wave front that illuminates the test antenna. This suppression is sought through one or more of such factors as

1. Directivity and sidelobe suppression of the source and test antenna
2. Clearance of the line of sight from the range surface
3. Redirection or absorption of energy reaching the range surface
4. Special signal processing techniques such as tagging by modulation of the desired signal, gating of the signals, and time-domain techniques or by use of short pulses.

Typical geometries associated with the free-space approach include the elevated range, the slant range, the rectangular anechoic chamber, the compact range, and above certain limiting frequencies, the tapered anechoic chamber.

Reflection ranges are designed to make use of energy that is reradiated from the range surface(s) to create constructive interference with the direct-path signal in the region about the test aperture. The geometry is controlled so that a symmetric small amplitude taper is produced in the illuminating field. The two major types of reflection ranges in use are the ground reflection range and, for low frequencies, the tapered anechoic chamber and the five-sided EMC emission chamber.

2.5.2 Electromagnetic Design Considerations and Criteria

For either basic type of range, the fundamental electromagnetic design criteria deal with control of

1. Inductive coupling between antennas
2. Phase curvature of the illuminating wave front

3. Amplitude taper of the illuminating wave front
4. Spatial periodic variations in the illuminating wave front caused by reflections
5. Interference from spurious radiating sources

Items 1 through 4 primarily establish the dimensional requirements on the range design as well as limiting the source-antenna directivity. Item 5 must be considered in the overall design.

2.5.3 Elevated Outdoor Antenna Range

Elevated antenna range design proceeds as follows (see Figure 2.10):

1. Determine the minimum range length from the range equation,

$$R = 2D^2/\lambda \tag{2-26}$$

2. Determine the maximum source antenna diameter where the amplitude taper is limited to 0.25 dB,

$$d \leq 0.37\ \lambda R/D \tag{2-27}$$

where
d = source diameter
D = test aperture
R = range length
λ = operating wavelength

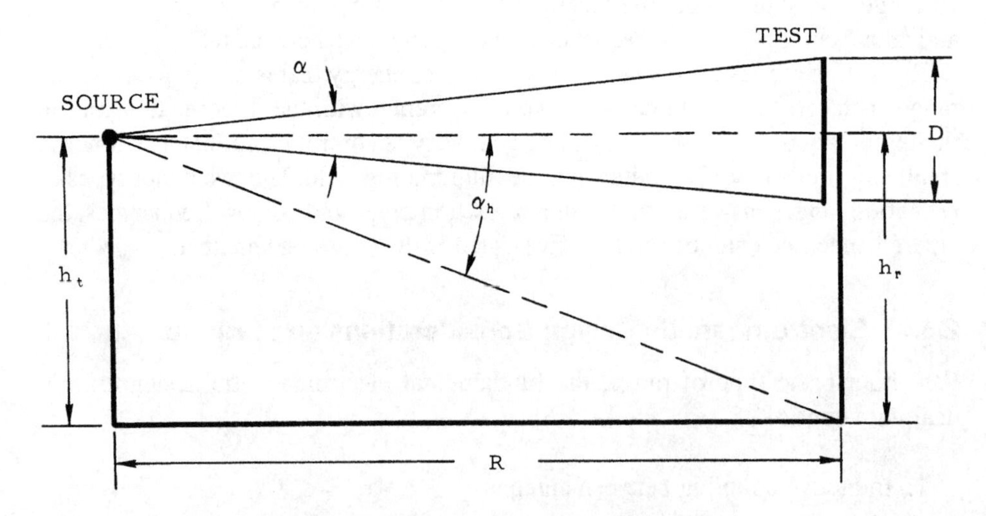

Figure 2.10 Elevated antenna range.

More generally, for a range length $R = KD^2/\lambda$, the source diameter would be restricted to

$$d \leq 0.37\, KD \tag{2-28}$$

Another criterion for elevated range design is given in Ref. 11, where

$$h_r \geq 4D$$

To ensure a good illumination taper and minimum range reflections, check to see that the first null of the source antenna pattern falls no lower than the base of the test antenna tower. If further reduction of range reflections is needed, this can be achieved by using range fences [11]. 2-14

2.5.4 Ground Reflection Antenna Range

2.5.4.1 Introduction. The ground reflection range design assumes that the energy from the source antenna cannot be kept from reaching the surface of the test range and reflecting into the test aperture. At a fixed frequency the range parameters can be adjusted so that the reflecting energy is phased so as to create constructive interference with the direct-path signal in the region of the test aperture.

2.5.4.2 Principles. The amplitude taper of the illuminating field along a horizontal line through the test aperture, normal to the line of sight, will be determined almost entirely by the source directivity, the aperture width, and the range length, just as for elevated ranges. The taper along a vertical line through the test aperture, however, is virtually independent of the directivity of the source antenna and depends almost entirely on the height of the center of the test aperture above the range surface. This may be shown as discussed in the following paragraphs.

Consider test-range geometry as sketched in Figure 2.11. Set the phase reference at zero for the direct-path wave at the source antenna and neglect the slight difference in loss due to dispersion along the path lengths for the direct and reflected waves. The phasor representing the amplitude of the field at points along a vertical through the test aperture can then be written as

$$E \cong E_D e^{-j\beta R_D} + kE_D e^{-j(\phi+\beta R_R)} \tag{2-29}$$

where the time dependence is suppressed. In this expression,

$$\beta = 2\pi/\lambda$$

$$R_D = [R^2 + (h - h_t)^2]^{1/2}$$

$$R_R = [R^2 + (h + h_t)^2]^{1/2}$$

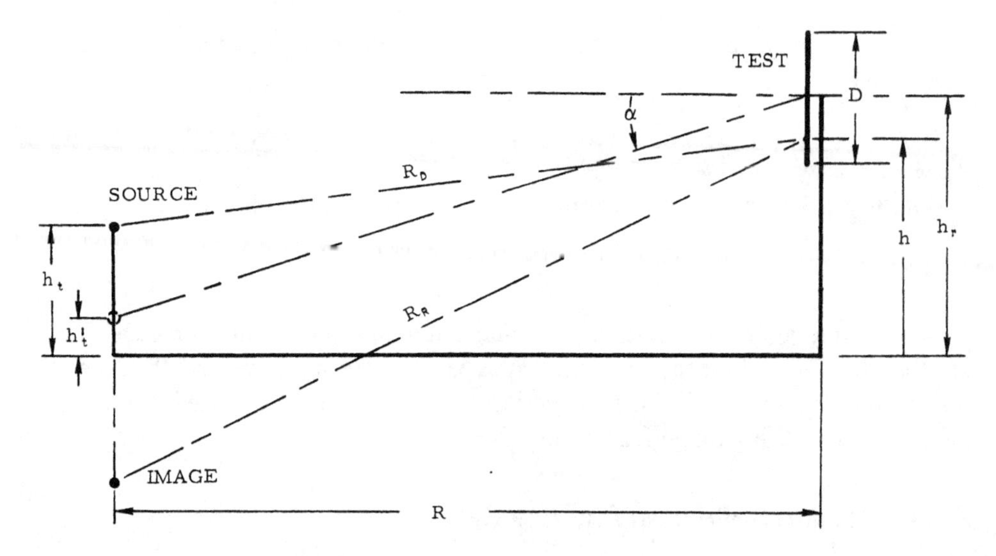

Figure 2.11 Ground reflection range.

E_D = the direct-path field amplitude
$ke^{-j\phi}$ = an effective reflection coefficient for the range surface
R_D = direct path length
R_R = reflection path length
R = range length
h_t = height of transmit antenna
h_r = height of receive antenna

Assuming that the phase shift for typical ground reflection geometry is nearly equal to π radians and that the range length R is large with respect to h_r, the height of the transmit source is given by approximately

$$h_t \cong \lambda R/(4h_r) \tag{2-30}$$

When the first interference lobe is peaked at the test height, the field variation is on the order of

$$E_n(h) \cong \sin(\pi h/2h_r) \tag{2-31}$$

To ensure that the vertical amplitude variation across the test aperture on a ground reflection range is sufficiently flat, it is recommended that the test height h_r be set at [11]

$$h_r \geq 3.3D \tag{2-32}$$

2.5.4.3 Control of Surface Reflections in the Ground-Reflection Mode. Reflections from a range surface may be classified into two general categories, diffuse scattering and specular reflection. The transition from conditions causing diffuse scattering to those producing essentially specular reflection is gradual. Accordingly, criteria for specifying the boundaries and tolerances for the reflecting surfaces and other controlled areas in the design of ground-reflection ranges are largely empirical.

For the range surface variations, it is suggested that peak-to-peak surface variation be held to within [11]

$$\nabla h \leq \lambda/(M \sin \varphi) \quad (2\text{-}33)$$

where

λ is the operating wavelength at the highest frequency of interest
M is a smoothness factor between 8 and 32
$\varphi = \arctan[(h_r + h_t)/R]$

For the amount of range surface area to be controlled, see the guidance given in Ref. 11.

2.5.5 Open-Area Test Sites (OATS)

2.5.5.1 Introduction. The Federal Communications Commission (FCC) was given the authority to impose rules and regulations on industrial, commercial, and consumer devices that may radiate electromagnetic energy under the Communications Act of 1934. The commission adopted the open-area test site (OATS), a specific form of ground reflection range, as a convenient method of standardizing emission measurements [17].

The original specification specified that radiated emissions were to be conducted on a ground plane using three range lengths: 3 m, 10 m, and 30 m. A suitable site required that it be clear of obstructions such as trees, bushes, or metal fences within an elliptical boundary. The site should also be well away from buildings, parked automobiles, and the like. The suitability of the site is determined by measuring the site attenuation [14].

2.5.5.2 The Design of Open Area Test Sites. Open-area test sites are ground-reflection ranges with specified range lengths and specified height ranges for the scanning antennas that are used to measure the radiated emissions. The result is that the sites have repeatable characteristics. Notably, the site attenuation, which is defined as the propagation loss between the item under test and the measurement antenna, becomes a quantifiable parameter. Given the site attenuation the level of radiated emissions from a device under test can be calculated. This is also true for free-space and anechoic chamber test sites.

The design of open-area test sites is now specified by ANSI C63.7-1992 [18]. An extensive discussion is given by Kodali [6].

The validity of the test site is determined by measuring the site attenuation as described in ANSI C63.4 [14].

REFERENCES

1. J. D. Kraus, and D. A. Fleisch, *Electromagnetics: With Applications*, McGraw-Hill, New York, 1998.
2. F. T. Ulaby, *Fundamentals of Applied Electromagnetics,* 1999 edition, Prentice-Hall, New York, 1998.
3. C. T. A. Johnk, *Engineering Electromagnetic Fields and Waves,* John Wiley & Sons, New York, 1975.
4. C. A. Balanis, *Engineering Electromagnetics,* John Wiley & Sons, New York, 1989.
5. IEEE Std 149-1979, *Standard Test Procedures for Antennas.*
6. V. P. Kodali, *Engineering Electomagnetic Compatibility,* 2nd edition, IEEE Press, 2001.
7. B. F. Lawrence, Anechoic Chambers, Past and Present, *Conformity,* Vol. 6, No. 4, pp. 54–56, April 2000.
8. MIL-STD-461E, *Requirements for the Control of Electromagnetic Interference Characteristics of Subsystems and Equipment.*
9. Special Session—Passive Intermodulation, *Antenna Measurement Techniques Association Proceedings,* pp. 142–163, 2000.
10. E. F. Knott, *Radar Cross Section Measurements,* Van Nostrand Reinhold, New York, 1993.
11. J. S. Hollis et al., *Microwave Antenna Measurements,* Scientific Atlanta, Inc., 1985 (available on CD from MI Technologies, Inc., Norcross, GA).
12. Ming-Kuei Hu, Fresnel Region Field Distributions of Circular Apertures, *IEEE Transactions on Antennas and Propagation,* Vol. AP-8, pp. 344–346, May 1960.
13. E. F. Knott et al., How Far is Far?, *IEEE Transactions on Antennas and Propagation,* Vol. AP-22, pp. 732–734, September 1974.
14. American National Standard Institute (ANSI), C63.4-2000: Standard for methods of measurement of radio noise emissions from low-voltage electrical and electronic equipment in the range of 9 kHz to 40 GHz.
15. European standard, EN61000-4-3:1996, Electromagnetic Compatibility (EMC) Part 4-3: Testing and measurement techniques—radiated, radio-frequency, electromagnetic field immunity test.
16. Harold T. Friis, A Note on a simple Transmission Formula, *Proceedings of the IRE,* Vol. 34, No. 5, p. 254, 1937.

17. Characteristics of Open Field Test Sites, *Bulletin OST 55,* Federal Communications Commission, Office of Science and Technology, Washington, D.C., August 1982.
18. ANSI C63.7-1992, American National Standard for Construction of Open-Area Test Sites for Performing Radiated Emission Measurements.

CHAPTER THREE

ELECTROMAGNETIC ABSORBING MATERIALS

3.1 INTRODUCTION

Electromagnetic absorbing material used in anechoic test facilities takes many forms, depending on the purpose of the facility and frequency of operation. The two most common are the dielectric absorbers used in the microwave frequency range and the ferrite absorbers used in the lower frequency range. Dielectric absorbers came about early in the development of anechoic chambers during the 1940s when indoor test facilities were first developed. These early chambers were lined with matted "horse hair" that was impregnated with a conductive carbon solution. In the 1950s, urethane foam was found to be a good carrier for the conductive solution, and the pyramidal-shaped absorber came into being. Well into the 1970s, this foam dielectric absorber was the primary product available for anechoic chamber applications. Then a series of serious chamber fires occurred which caused the government to research the fire retardant properties of absorbers and develop a new fire safety standard. The foam absorbers now marketed must meet the requirements of Naval Research Laboratory (NRL) 8093 [1]. This document specifies a series of laboratory tests to be performed to provide a grade of absorber that is quite fire-resistant. For more details on the history of anechoic chambers, refer to Refs. 2 and 3. In the late 1980s, a market developed for low-frequency testing within an indoor chamber. The first of these facilities used the then available material, which were large absorber pyramids installed within shielded enclosures to establish an acceptable testing environment. These facilities were very expensive, and only the larger companies could support the cost. As the demand for facilities rose, other solutions were developed, and the ferrite-lined chamber came into commercial use. Today there are several sources for the ferrite material. Most suppliers of anechoic facilities provide designs for the full range of materials. There are thousands of these chambers worldwide. A detailed history of EMI measurements is given in Chapter 1 of Ref. 4. In the 1990s, chambers for testing wireless devices became the vogue, and a variety of new chamber designs were developed [5]. All types of absorbing material used in anechoic chambers are reviewed in this chapter. For convenience in presentation, the section is broken down into two classes of materials, those that are used above 1 GHz (microwave) and those used below 1 GHz (low frequency).

3.2 MICROWAVE ABSORBING MATERIALS

3.2.1 Pyramidal Absorber

3.2.1.1 Solid Foam. Most of the electromagnetic anechoic chamber manufacturers offer a standard microwave absorber product that is pyramidal in shape. The product is a solid carbon-loaded urethane foam absorber. To meet the current fire retardancy requirements it is loaded with fire-retardant chemicals that are either mixed in with the carbon solution or added as a second treatment. The primary application for this material is in the construction of anechoic chambers, or for covering test equipment within the chambers. Of all known absorbers, it provides the highest broadband performance at both normal incidence and at wide incidence angles. It is primarily used for reducing forward scattering, but offers good back scattering properties as well, making it suitable for use in all locations within an anechoic chamber. It is available in a variety of thicknesses. It provides the chamber designers with the opportunity to choose the right product for specific frequencies and incidence angles.

The industry provides material as small as 5.1 cm (2 in.) high and as thick as 3.7 m (12 ft). The product generally comes in standard thicknesses. Special cuts are often made for specific applications. Table 3.1 summaries common pyramidal materials available from the microwave materials industry. Two geometries are available; the straight square pyramid shown in Figure 3.1, and the twisted pyramid illustrated in Figure 3.2. The latter is used for the larger pyramids where the shoulders of the pyramids help carry the load of the cantilevered geometry, illustrated in Figure 3.3. The twisted pyramid is commonly used in the backwall of tapered chambers where the absorbers commonly used are up to 1.8 m (6 ft) thick.

The product is black when made and is usually painted with a blue latex paint to improve light reflectance. At 95 GHz, the paint can degrade absorber reflectivity as much as 5 dB, so the tips are often left unpainted in chambers where millimeter measurements are to be conducted. It also has been found that if the tips are left unpainted, it minimizes tip breakage due to wear and tear found in normal chamber operations. Foam absorber is not very robust and requires a high degree

Table 3.1 Pyramidal Absorber Performance

Type	Height, cm (in.)	Weight, kg (lb)	Tips per piece	Normal Incidence Reflectivity, GHz									
				0.12	0.3	0.5	1.0	3.0	6.0	10.0	18.0	36	50
P-4	10.9(4.3)	1.4(3)	144					30	35	42	50	50	50
P-6	15.2(6)	1.6(3.5)	100					32	40	45	50	50	50
P-8	20.3(8)	2.0(4.5)	64				30	37	45	50	50	50	50
P-12	30.5(12)	2.7(6.0)	36				35	40	45	50	50	50	50
P-18	45.7(18)	5.4(12)	16			30	37	40	45	50	50	50	>45
P-24	61(24)	7.7(17)	9		30	35	40	45	50	50	50	50	>45
P-36	91.4(36)	10.9(24)	4		35	37	42	50	50	50	50	50	>45
P-48	121.9(48)	17(38)	2	28	35	40	50	50	50	50	50	50	>45
P-72	182.9(72)	23(50)	1	33	40	45	50	50	50	50	50	50	>45

Note: Base dimensions are 0.61 m^2(2 ft^2). Power rating is 0.08 W/cm^2(0.5 W/$in.^2$).

of maintenance. This is especially true in the vicinity of doors and other high-traffic areas.

Foam absorbers are generally installed using pressure-sensitive adhesives. The chamber surface is first painted with the adhesive, and then the back of the absorber is coated. The adhesive is left to flash off, and then the block is applied to the surface. Several adhesive products on the market are used in these types of installations. It is best to use the adhesives recommended by the absorber manufacturer, because they have confirmed the compatibility of the adhesive with the chemistry of the absorber material. If not properly applied, the adhesives tend to lose their bonding power and the pyramids start to come loose from the walls and especially the ceiling.

Figure 3.1 Geometry of the standard pyramid absorber. (Photograph courtesy of Advanced ElectroMagnetics, Inc., Santee, CA.)

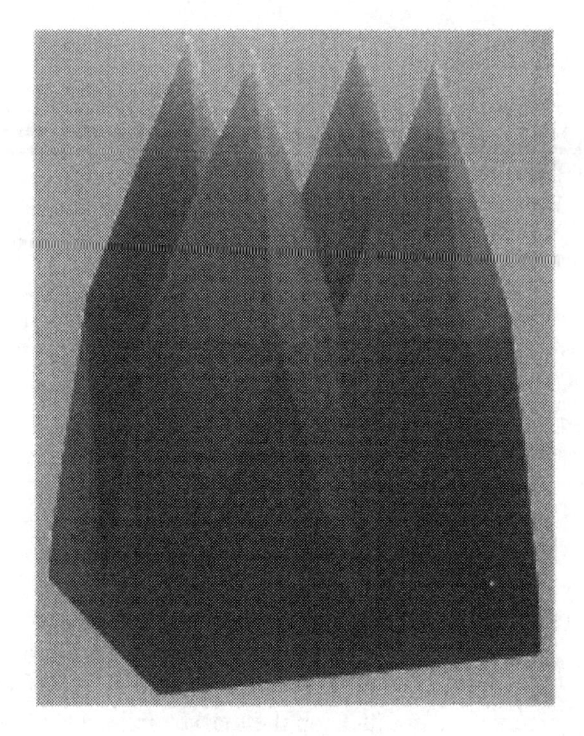

Figure 3.2 Geometry of the twisted pyramid absorber. (Photograph courtesy of Advanced ElectroMagnetics, Inc., Santee, CA.)

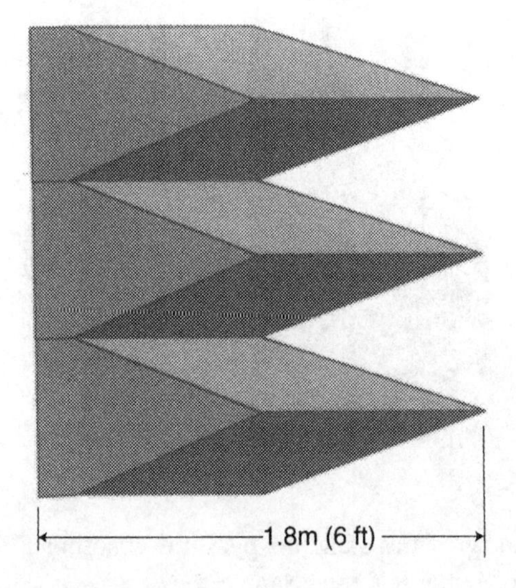

Figure 3.3 Stacking of twisted pyramidal absorber.

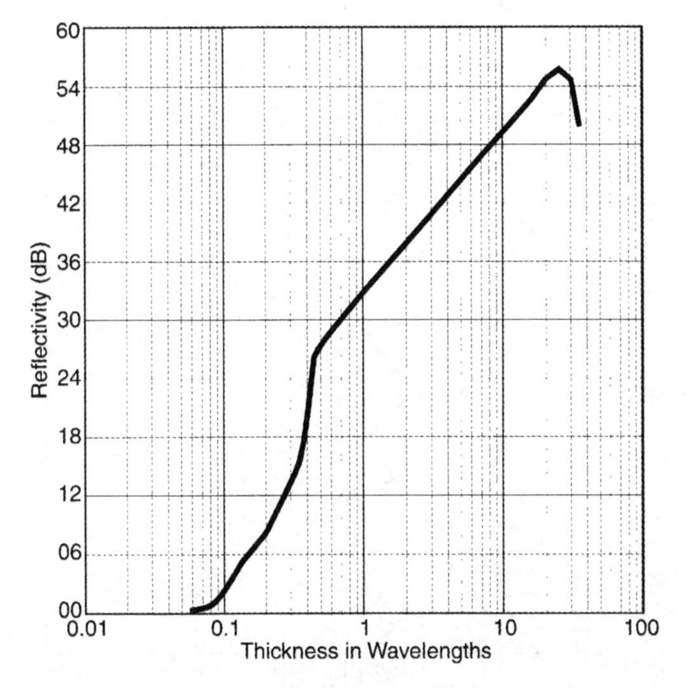

Figure 3.4 Normal incidence reflectivity performance of pyramidal absorbers.

Pyramidal electromagnetic performance is specified as reflectivity at normal incidence and is stated in –dB. Generally, this information is provided in tables of absorber thickness versus frequency (see Table 3.1 [6]) over the decibel range of –30 to –50. When the loading is optimized and the material is on the order of one wavelength thick, the reflectivity is about –33 dB. At eight wavelengths, the reflectivity is on the order of –51 dB. The numbers in the manufacturer's charts are rounded off for convenience in specifying the products. Figure 3.4 shows typical performance of pyramidal materials.* The slope of the curve is approximately 6 dB per octave, the same as a one-section filter. It is also the same loss that is experienced in free space by a wave front when the distance or frequency is doubled. When the absorber is less then one wavelength thick, the performance rolls off and approaches 0 dB very quickly. When the absorber is thick and large in terms of wavelengths, the incident wave front begins to scatter and the absorber reflectivity performance drops off. The actual drop-off is a function of the aspect ratio (height/base dimensions) and the nature of the conductive loading used in the product to achieve the microwave loss properties. Experience has shown that the curve in Figure 3.4 is achieved when the conductivity of the carbon loading is optimal. If the conductivity is too great (i.e., the carbon loading is high), then the reflectivity drops at the low end of the curve and peaks at the higher thicknesses as shown in Figure 3.5. If the loading is too light, then the whole curve shifts to

*The curve was developed by the author in 1975 from published absorber performance tables. The curve has been validated in many measurements of pyramidal absorber materials.

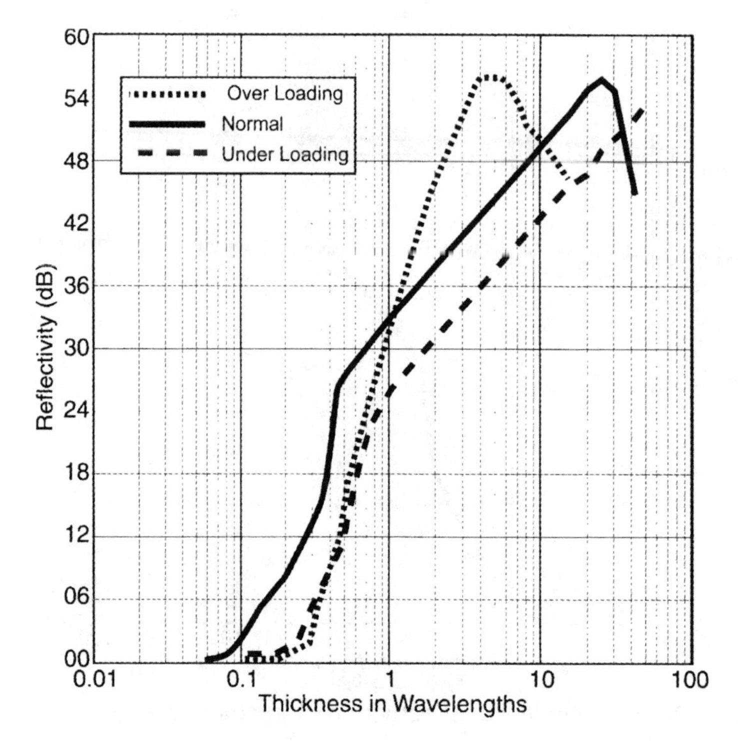

Figure 3.5 Effect of carbon loading variations in the performance of pyramidal absorbers.

the right. It takes a thicker material to achieve the same performance. Thus, it is best to control the properties of the material using a test procedure that checks the reflectivity of the absorber when it is on the order of one wavelength thick. Another important property of the electromagnetic performance is the reflectivity at wide angles of incidence. This is especially important for determining the forward scattering performance in anechoic chambers. The industry has developed a graph (Figure 3.6*) of bistatic absorber performance where the angle of incidence is plotted versus the absorber thickness in wavelengths. It is very common for the aspect ratio (length/width) of an anechoic chamber to cause the angle of incidence of the wave front to fall to the 60 to 70-degree range. As can be seen, the absorber performance drops off very rapidly versus angle of incidence, illustrated in Figure 3.7. Figure 3.6 is used in calculating the level of reflected energy that will be found in the test region of the chamber due to reflections from the sidewall(s), ceiling, and floor.

A couple of absorber design improvements have recently been reported [7, 8]. The first of these improves the low-frequency performance and is called the doubly periodic curved absorber. The pyramidal material is cut such that the base of the pyramid is extended up about one-half the length of the unit and then curves

*See discussion on bistatic performance in reference 2.

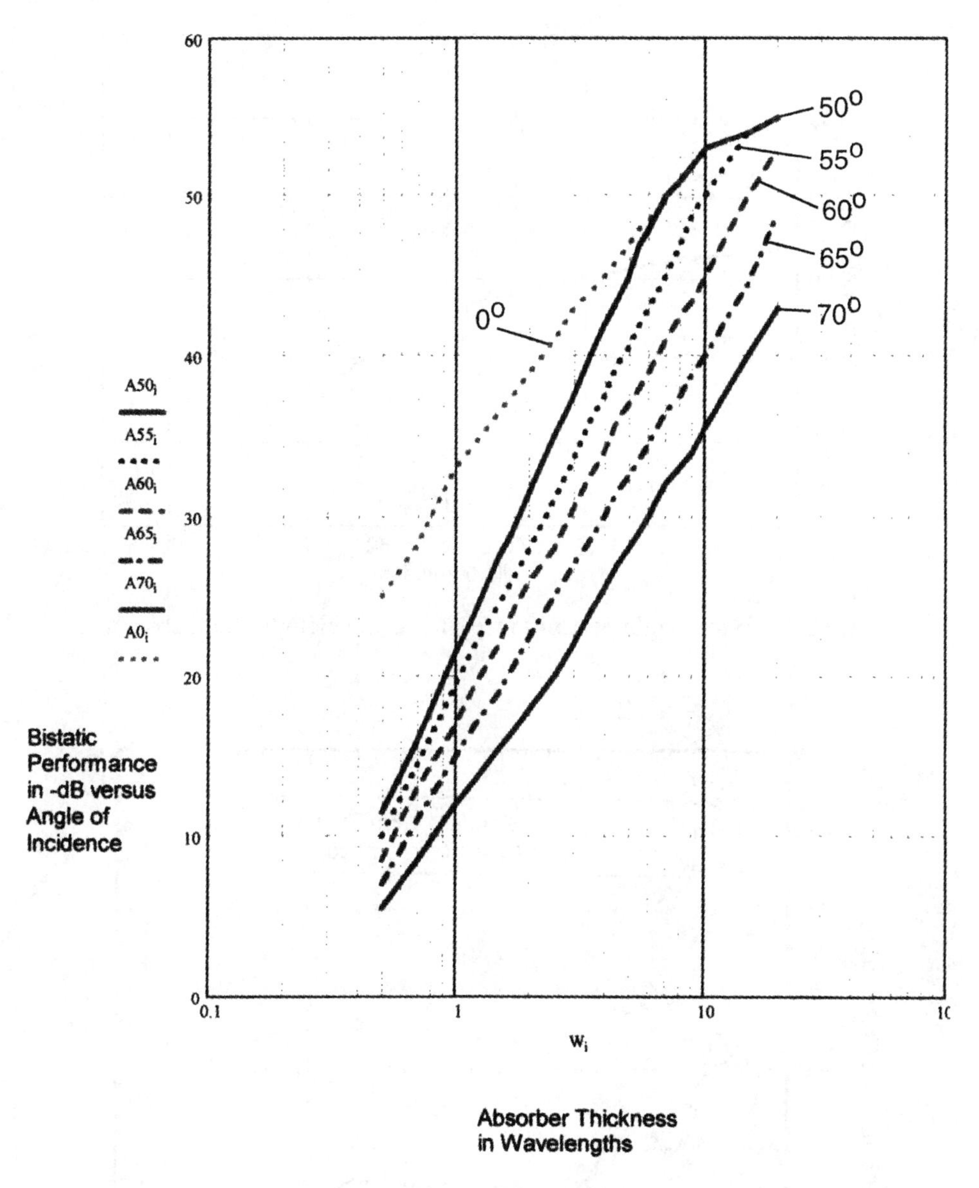

Figure 3.6 Wide-angle (bistatic) performance of pyramidal absorbers.

into a sharp point at the top. This provides more volume for the lossy material. For a given height, the absorber performance is extended down in frequency, effectively reducing the height for a given frequency band. Two versions were developed. The first was 1 m (40 in.) high and operates down to 300 MHz, and the second is 1.52 m (60 in.) high and operates down to 200 MHz. The performance of the 1.0-m (40-in.) design is compared to a standard 1.82-m (72-in.) pyramid in Figure 3.8. The second improvement was achieved by using a Chebychev distribution, wherein the pyramids were varied in length to break up the periodic

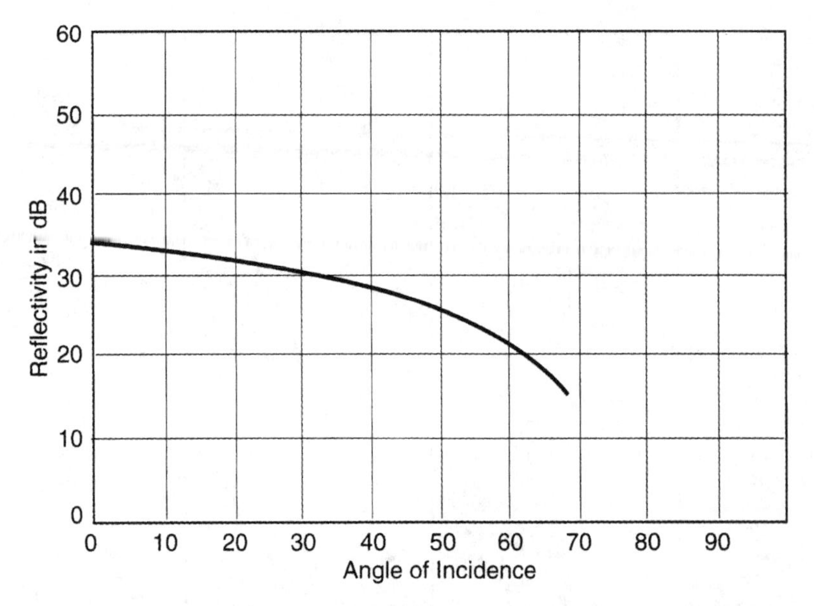

Figure 3.7 Wide-angle performance of one wavelength thick absorber.

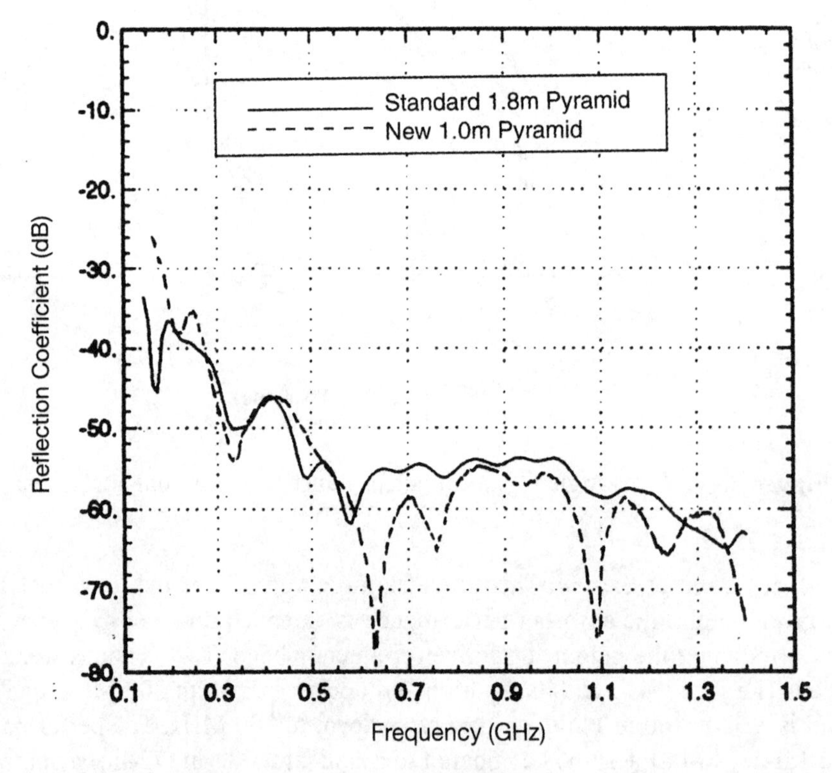

Figure 3.8 Performance comparison of standard 1.82-m to 1.0-m doubly periodic curved pyramidal absorber.

arrangement of the pyramids. This random phasing of the pyramids added another 10 dB of reflection performance compared to a standard pyramid of the same height. The Chebychev design has been applied to both the standard and the doubly periodic curved pyramidal materials, with excellent results in both cases. The tradeoff is reduced pyramidal height for increased cost of manufacture for a given performance level at normal incidence. Analysis has shown that the Chebychev reduction factor decreases with bistatic angle of incidence due to the shrinking effective size. The concept is best suited for near-normal incidence [8].

3.2.1.2 Hollow Pyramidal Absorber. Two forms of hollow pyramidal absorber are available. The most common is a product made by taking thin flat sheets of foam absorber and wrapping them around a square or round lightweight core that has the pyramidal geometry. Both multilayer and single-layer foam has been used to form this type of product. The individual cones are grouped into 0.61-m (2-ft) squares on a common base to make up a convenient unit for chamber installation. The height of this absorber is similar to that of the solid pyramidal absorber. This product has been extensively used in Europe.

Another form of this design is a product [9] that is made by spraying plastic film with a thin metallic material, which is then bonded to pyramidal foam supports. This design is illustrated in Figure 3.9. Both approaches provide good fire retardancy and lightweight installation. These types of material are especially useful for operation below 1 GHz, where the pyramidal materials must be long in length to provide good performance down to 30 MHz.

3.2.2 Wedge Absorber

Another form of the foam absorber is wedge material. This product has the same geometry as the pyramidal material in one direction and has uniform shape in the other, as illustrated in Figure 3.10. It is used in chamber designs where it is desirable to have the energy guided into a terminating wall, such as in tapered chambers and in compact ranges designed for RCS measurements. The reflectivity is the same as the pyramidal materials of the same height, where the electric field is perpendicular to the wedges and about 10 dB less when copolarized with the direction of the wedges (when the absorber is on the order of eight wavelengths thick). The new double curvature and Chebychev designs have also been applied to the wedge absorber with similar improvement in lowering the frequency of operation for a given thickness of material. An application of these materials is described in the section on the tapered chamber.

3.2.3 Convoluted Microwave Absorber

Another broadband microwave absorber is made using convoluted foam (see Figure 3.11). This manufacturing technique is used in the furniture industry, and it found early use in the microwave industry. The product is still available and

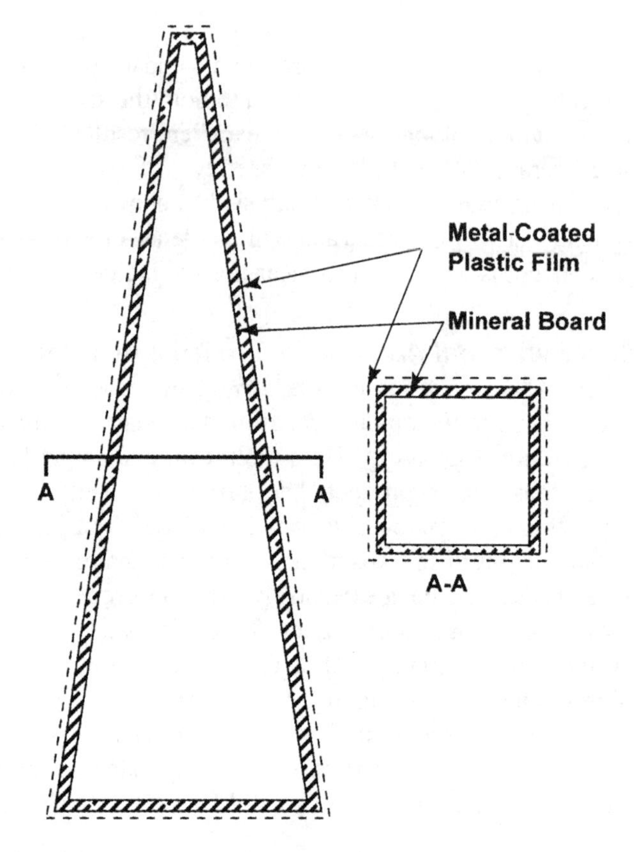

Figure 3.9 Construction of a metal film absorber.

Figure 3.10 Geometry of wedge-shaped absorber material. (Photograph courtesy of Advanced ElectroMagnetics, Inc., Santee, CA.)

Figure 3.11 Geometry of the convoluted absorber. (Photograph courtesy of Advanced ElectroMagnetics, Inc., Santee, CA.)

comes in various grades (thicknesses). It is primarily useful in the upper microwave frequency ranges, especially in the millimeter bands. An example of the products available and their performance is given in Table 3.2 [10].

3.2.4 Multilayer Dielectric Absorber

Multilayer dielectric absorber is formed from sheets of uniform treated foam. The layer thickness and carbon loading vary, depending on the overall thickness of the product. A family of products, as illustrated in Table 3.3, is available. They vary, depending on the desired frequency coverage. The thicker the product, the lower the frequency of operation of the material, as illustrated in Figure 3.12 [11]. This product is used in many laboratory applications. The theory describing the operation and design of these types of absorbers can be found in Ref. 14.

3.2.5 Hybrid Dielectric Absorber

Multilayer absorber is very efficient in terms of providing loss in a minimum of height. It provides 33% more dielectric material then a pyramidal absorber of the

Table 3.2 Typical Performance of a Convoluted Microwave Absorber

Type	Height, cm (in.)	Normal Incidence Performance, GHz					Power Rating, W/cm^2 (W/in.^{2})
		3.0	6.0	10.0	15.0	30.0	
C-1.5	3.8(1.5)		20	30	35	45	0.08 (0.5)
C-3	7.6(3.0)	20	30	40	45	50	0.08 (0.5)
C-4	10.1(4.0)	25	20	30	35	45	0.08 (0.5)

Note: Base dimensions are 0.61 m^2 (2 ft^2).

Table 3.3 Properties of the Multilayer Absorber

Type	Range Frequency	Bands Covered	Maximum Power Reflectivity	Sheet Size	Nominal Thickness		Weight	
							kg/m²	lb/ft²
AN72	20 GHz and above	K	1%	61 cm × 61 cm (24 in. × 24 in.)	0.6 cm	$\frac{1}{4}$ in.	0.5	0.1
AN73	7.5 GHz and above	X, Ku, K	1%	61 cm × 61 cm (24 in. × 24 in.)	1.0 cm	$\frac{3}{8}$ in.	1.0	0.2
AN74	3.5 GHz and above	C, A, B, X, Ku, K	1%	61 cm × 61 cm (24 in. × 24 in.)	1.9 cm	$\frac{3}{4}$ in.	1.5	0.3
AN75	2.4 GHz and above	S, C, A, B, X, Ku, K	1%	61 cm × 61 cm (24 in. × 24 in.)	2.9 cm	$1\frac{1}{8}$ in.	2.4	0.5
AN77	1.2 GHz and above	L, S, C, A, B, X, Ku, K	1%	61 cm × 61 cm (24 in. × 24 in.)	5.7 cm	$2\frac{1}{4}$ in.	4.4	0.9
AN79	0.6 GHz and above	L, S, C, A, B, X, Ku, K, UHF	1%	61 cm × 61 cm (24 in. × 24 in.)	11.4 cm	$4\frac{1}{2}$ in.	9.8	2.0

same thickness. Thus a combination of the two approaches is useful in making a low-frequency absorber for applications below 1 GHz. These dielectric hybrids consist of a pyramidal front material properly loaded to match successive layers of graded dielectric layers. The pyramidal front provides for a tapered front face impedance match passing energy into successively higher dielectrically loaded materials. This enhances energy dissipation. It is through this gradual change in dissi-

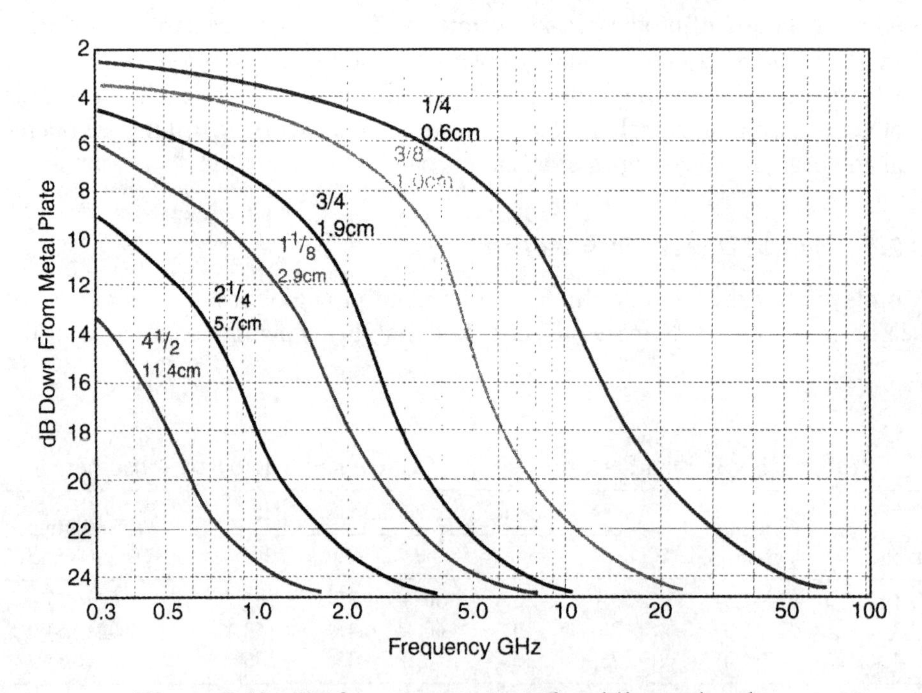

Figure 3.12 Performance curves of multilayer absorber.

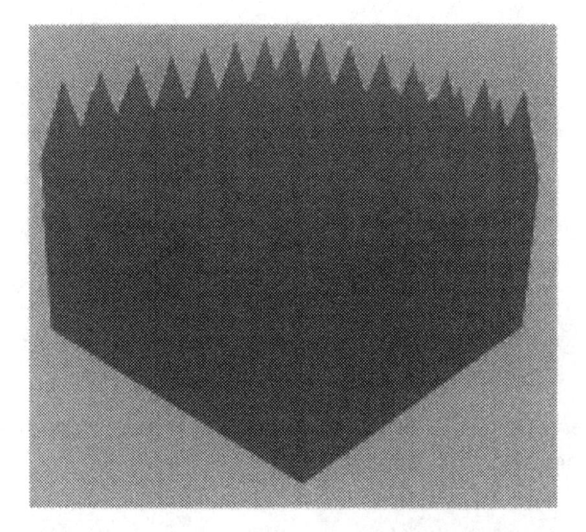

Figure 3.13 Geometry of hybrid dielectric absorber. (Photograph courtesy of Advanced ElectroMagnetics, Inc., Santee, CA.)

pation that high loss at low frequencies can be achieved. Due to the large square base of the product (Figure 3.13), it permits the material to be easily stacked, one on top of another. Table 3.4 provides typical performance of this product [13]. Typical performance increases at normal incidence are 3–6 dB, and 6–10 dB at wide angles, when compared to a pyramidal absorber of the same height.

3.2.6 Walkway Absorber

Broadband walkway absorber is manufactured by placing self-extinguishing polystyrene core in an interlocking pattern over the top of a standard pyramidal absorber. This is then topped with a higher density polystyrene laminate. This construction is illustrated in Figure 3.14. Typical mechanical and electrical specifications are given in Table 3.5 [14].

Table 3.4 Performance Data of the Hybrid Dielectric Absorber

		Typical Absorber Performance at Normal Incidence, MHz					
Type	Height, m (in.)	25	50	100	250	500	1000
P-12EM	0.3(12)	2	4	9	19	26	32
P-18EM	0.46(18)	3	6	11	21	29	35
P-24EM	0.61(24)	4	8	14	26	32	38
P-36EM	0.92(36)	6	12	21	30	36	41
P-48EM	1.22(48)	9	16	27	34	39	44

Note: Above 1000 MHz the performance of the EM series is essentially the same as the standard pyramidal material. The base dimensions are 0.61 m^2 (24 $in.^2$).

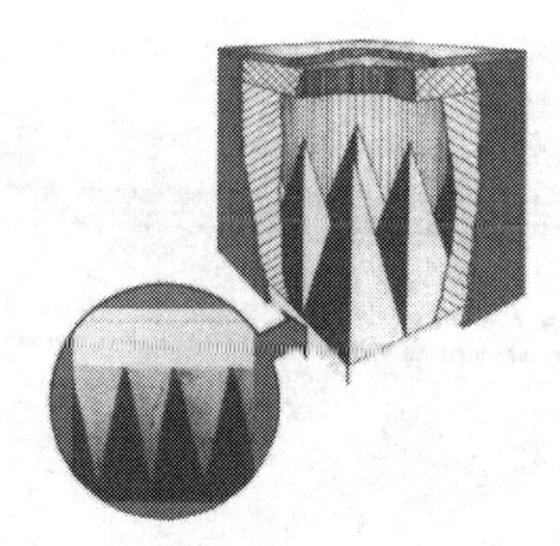

Figure 3.14 Geometry of a walkway absorber. (Photograph Courtesy of Advanced ElectroMagnetics, Inc. Santee, CA.)

3.3 LOW-FREQUENCY ABSORBING MATERIAL

3.3.1 Introduction

The large demand for testing in the 30- to 1000-MHz frequency range has brought about the development of a series of materials optimized for this application. The most common material is ferrite tile. Ferrite tile has been used for low-frequency applications for over 40 years. The increase in shielded enclosures developed a volume market for the material. Several sources and products are now available.

The tile used for lining anechoic chambers comes in two geometries. The most common is a flat ceramic tile about 6 mm thick and 100 mm^2. The other version is in the form of a cast ceramic grid. The early versions of the grid were on the order of 1 in. thick. A newer grid version is on the order of one-half inch thick. The material is very heavy, and is normally installed onto plywood panels attached to the walls of the shielded enclosure. Thus, the overall weight of the installation is considerably greater than that of a foam installation. A bonus is that installations are extremely fire-retardant, and a simple fire suppression system can be used.

Table 3.5 Typical Specifications for the Walkway Absorber

Type	Thickness, cm (in.)	Load Rating, kg/m^2 (lb/ft^2)	Reflectivity Performance at Normal Incidence, GHz							
			0.25	0.50	1.0	3.0	6.0	10.0	15.0	30.0
WW-4	14.0(5.5)	61.5 (300)				20	25	30	35	40
WW-8	23.3(9.5)	61.5 (300)			20	27	35	40	40	40
WW-12	36.8(14.5)	61.5 (300)			20	30	35	40	40	40
WW-18	54.6(21.5)	82.0 (400)		20	25	30	35	40	40	40
WW-24	61.0(29.5)	82.0 (400)	20	25	30	35	40	40	40	40

Note: Power rating is 0.08 watts/cm^2(0.5 watts/in.^{2}). Base dimensions are 0.61 m^2 (2 ft^2).

A new product recently introduced is a low-frequency adaptation of a high-frequency design. The aerospace industry has been using honeycomb radar-absorbing material since the 1950s. Of a similar construction, a conductive cellular structure (CCS) absorber is now available for use in low-frequency chambers [15]. It is a cell-based product where the cells are covered with an absorbing coating. The depth of the cells are chosen so that the incident wave sees the reflected signal to be 180 degrees out of phase, and the summation is a signal attenuated by about 20 dB. Broadbanding is achieved by nesting smaller cells within the larger cells and stagger-tuning the set. Typical units are on the order of 0.61 m^2 (2 ft^2) and about 0.3 m (1 ft) high. The product is made from materials that have been treated with fire-retardant chemicals. When the CCS absorber is topped with pyramidal material, the performance is enhanced throughout the whole band of operation and is especially useful in extending the performance up into the microwave range [16].

3.3.2 Ferrite Absorbers

Ferrite absorbers, illustrated in Figure 3.15, provide an alternative to the traditional large, foam-type absorber materials. They save chamber volume and reduce the fire hazard. Ferrite absorbers are inherently immune to fire, humidity, and chemicals. They provide a reliable and compact solution for attenuating plane wave reflections in shielded chambers in the 30- to 1000-MHz frequency band.

The basic physics of operation for any planar electromagnetic absorber involves the fundamental concepts depicted in Figure 3.16. When an electromag-

Figure 3.15 Flat ferrite tile absorber is available in 100-mm^2 units.

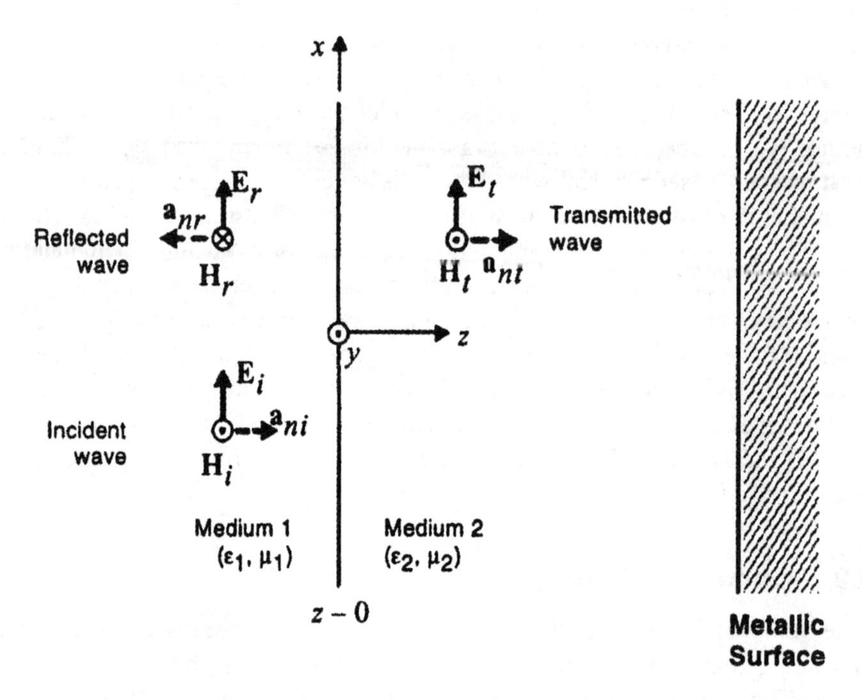

Figure 3.16 The physics of planar ferrite tile operation.

netic wave traveling through free space encounters a different medium (at $Z = 0$), the wave will be reflected, transmitted, and/or absorbed. It is the magnitude of the reflected signal that is of interest in anechoic chambers. For ferrite tiles, the thickness is adjusted so that the relative phases of the reflected and exiting wave cancel to form a broad resonant loss condition. The resonant condition appears as a deep "null" in the return loss response. This resonance is also a function of the frequency-dependent electrical properties of the ferrite material, such as relative permeability (u_r) and permittivity (ε_r) that interact to determine the reflection coefficient (Γ), impedance (Z), and return loss (RL) [17].

For most chamber applications, increased absorber bandwidth may be required. One method is to mount the tile over a dielectric spacer (typically wood) of various thicknesses. When both tile and spacer thickness is optimized, the frequency response is shifted upward to improve the high-frequency response. The performance of different designs are shown in Figure 5.17 [18]. In chamber design, the wide-angle performance of absorber is important. This is illustrated for ferrite tile in Figure 5.18.

The return loss is a function of the gaps between the tiles. Most tiles are precision machined to 0.005-in. tolerances on all surfaces. Care must be taken to minimize gaps between tiles in the actual installation. Most ferrite tiles come in 100-mm^2 units. At least one source provides units in 200-mm^2 units, and this reduces the gap problem significantly.

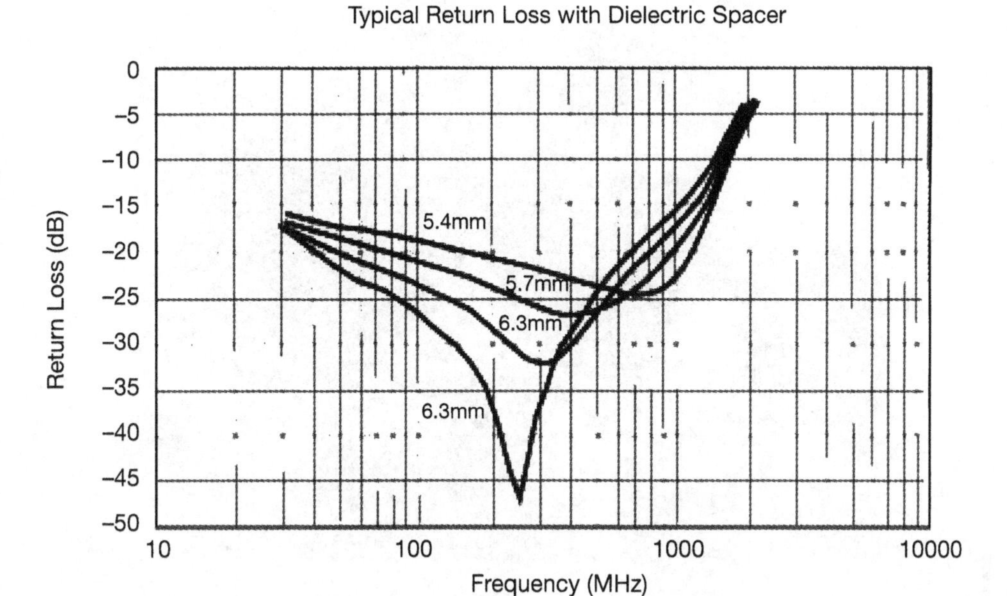

Figure 3.17 Planar ferrite tile performance.

Ferrite grid absorbers are ceramic castings, shown in Figure 3.19. The air–ferrite interface broadens the frequency response of the absorber but requires the material to be thicker to achieve the same performance as the solid ferrite material. It is also more tolerant to gaps between the tiles than the solid material. Typical performance of the grid tile absorbers is shown in Figure 3.20 [18].

The solid and grid ferrite tile absorbers have been successfully used in 3-m

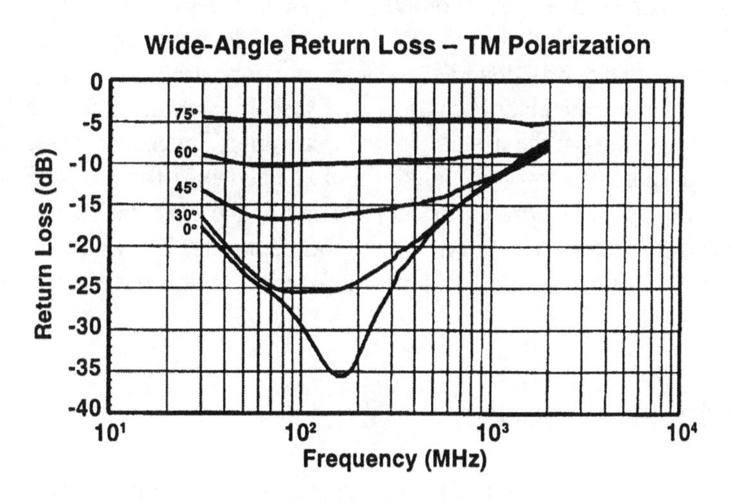

Figure 3.18 Wide-angle performance of planar ferrite tile.

Figure 3.19 Geometry of ferrite grid absorber. (Photograph courtesy of Panashield, Inc., Norwalk, CT.)

chambers for emission and immunity testing over the 30- to 1000-MHz frequency band.

3.3.3 Hybrid Absorbers

Hybrid absorbers combining both ferrite and dielectric absorbers have taken many forms. One of the most widely used material combinations is of flat ferrite tile and lossy foam wedge. The geometry is shown in Figure 3.21. Normal incidence performance is included in Figure 3.22 [19]. The marriage of the two materials provides for a very broadband absorber that can be used from 30 MHz to beyond 18 GHz in anechoic chambers used for EMC. It is common for the absorber supplier to custom design the material for the customer's needs. Contact with various material suppliers is recommended, should a need exist for this type of chamber design. The figure illustrates the normal incidence performance of some of the material available from a number of manufacturers. These absorbers have been found to provide good performance in 10-m emission measurement chambers.

3.4 ABSORBER MODELING

In recent years, since the advent of low-frequency testing in anechoic chambers, it has become necessary to develop computer models of absorbing materials. Per-

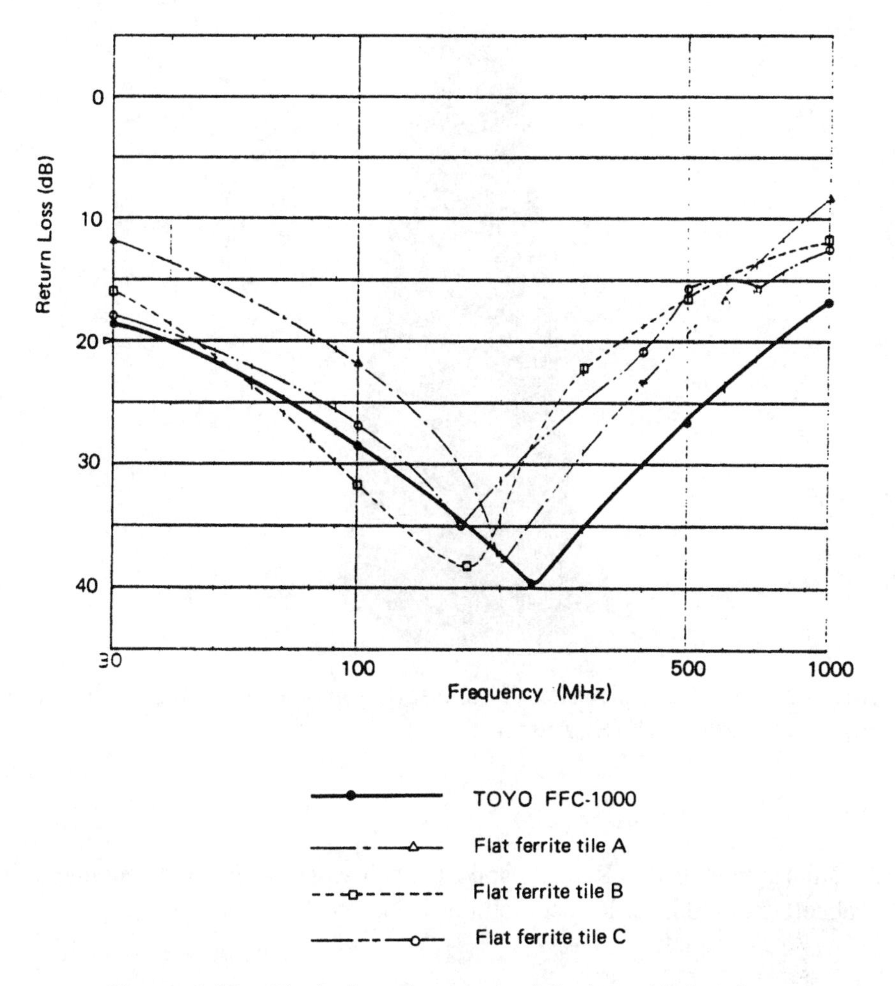

Figure 3.20 Typical performance of ferrite grid absorber.

haps the most documented work has been done at the University of Colorado at Boulder. They developed a method of evaluating the interaction of electromagnetic waves using an array of absorbing wedges or pyramid cones down to the low-frequency limit—that is, where the period of the array is small compared to the wavelength. A theoretical model is obtained using a method of homogenization. This replaces the transversely periodic structure with a transversely uniform medium possessing a certain (generally anisotropic) effective permittivity and permeability. Plane-wave reflection from such structures can then be modeled using well-known techniques for one-dimensionally inhomogeneous media. A Riccati equation for the reflection coefficient is used in this work. This model is appropriate for use with absorber found in anechoic chambers used for electromagnetic compatibility and electromagnetic interference (EMC/EMI) measurements over the frequency range of 30–1000 MHz. Their approach and

Figure 3.21 Ferrite/Dielectric hybrid absorber geometry. [Photograph courtesy of EMC Test Systems (ETS), Austin, TX.]

test results are detailed in Refs. 20 and 21. Their work has led to a new line of hybrid absorbers used in EMC test facilities [22].

3.5 ABSORBER TESTING

It is vitally important that absorbing material be capable of being measured prior to installation in an anechoic chamber. The materials are expensive, and the installation labor costs are very high. The industry standard has been the use of what is known as the NRL arch [2]. This consists of a pair of horns located at near normal incidence and focused toward a plate the size of the absorber sample, as illustrated in Figure 3.23. The plate is used as a reference. The absorber is then placed upon the plate, and the difference in the received signal is the reflectivity of the absorber. This works very well for finite size samples up to about 3.7 m^2 (12 ft^2). At the low frequencies it is not practical. Waveguide and rectangular coaxial techniques are used down to the 30-MHz frequency range. These techniques are discussed in Ref. 23.

It is highly recommended that all dielectric absorbing materials be 100% tested prior to delivery. As noted in the section on absorber performance, it is important that the test be performed when the material is on the order of a wavelength thick

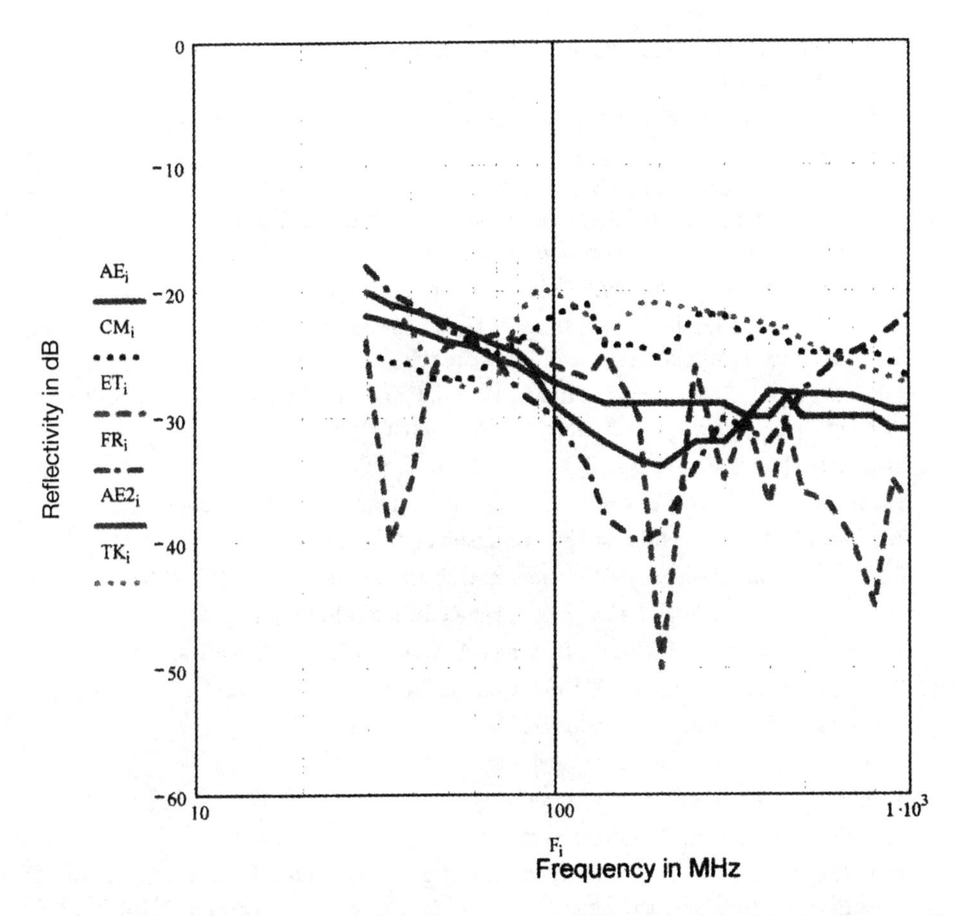

Figure 3.22 Performance of various magnetic/electric hybrid absorbers. AE refers to Advanced Electromagnetics, Inc., CM refers to Cuming Microwave Corp., and ET refers to Electromagnetic Test Systems (ETS.)

for microwave materials. Sample lots of the hybrid ferrite material should be tested to ensure that the material meets their published specifications.

In addition to performance testing, it is also necessary to conduct flame resistance testing. A variety of these tests have evolved. They are reviewed in Chapter 9.

REFERENCES

1. P. A. Tatem et al., *Modified Smoldering Test of Urethane Foam Used in Anechoic Chambers,* NRL Report 8093.
2. W. H. Emerson, Electromagnetic Wave Absorbers and Anechoic Chambers Through the Years, *IEEE Transactions on Antennas and Propagation,* Vol. AP-21, No. 4, p. 484, July 1973.

3. B. F. Lawrence, Anechoic Chambers, Past and Present, *Conformity,* Vol. 6, No. 4, pp. 54–56, April 2001.
4. V. P. Kodali, *Engineering Electromagnetic Compatibility,* 2nd edition, IEEE Press, New York, 2001.
5. J. Krogerus, Anechoic Chamber with Easily Removable 3D Radiation Pattern Measurement System for Wireless Communication Antennas, *Antenna Measurement Techniques Association Proceedings,* p. 31, 2000.
6. Technical Data Sheet, *Pyramidal Absorber,* Cuming Microwave.
7. I. J. Gupta et al., Design and Testing of New Curved Pyramidal Absorber, *Antenna Measurement Techniques Association Proceeding,* p. 50, 1997.
8. J. R. Gau et al., Chebychev Multilevel Design Concept, *IEEE Transactions of Antennas and Propagation,* Vol. AP-45, p. 1286, August 1997.
9. Technical Data Sheet, *Hollow Pyramidal Absorber,* Frankonia.
10. Technical Data Sheet, *Convolute Absorber,* Advanced ElectroMagnetics, Inc.
11. Technical Data Sheet, *AN Absorber,* Emerson and Cumings.
12. E. F. Knott et al., Radar Cross Section, Artech House, Boston, p. 247, 1985.
13. Technical Data Sheet, *EM Absorber,* Advanced ElectroMagnetics, Inc.
14. Technical Data Sheet, *Walkway Absorber,* Advanced ElectroMagnetics, Inc.
15. Patent No. 312,798, March 10, 1964, *Conductive Cellular Structure Absorber.*
16. Technical Data Sheet, *CCS Absorbing Material,* AEMI, 2000.
17. Technical Data Sheet, Fair-Rite Corporation.
18. Technical Data Sheet, Toyo Corporation.
19. Technical Data Sheet, EMC Test Systems.
20. E. F. Kuester, and C. L. Holloway, A Low-Frequency Model for Wedge or Pyramid Absorber Arrays-I: Theory, *Transactions on Electromagnetic Compatibility,* Vol. 36, No. 4, p. 300, November 1994.
21. C. L. Holloway and E. F. Kuester, A Low-Frequency Model for Wedge or Pyramid Absorber Arrays—II: Computed and Measured Results, *Transactions on Electromagnetic Compatibility,* Vol. 36, No. 4, p. 307, November 1994.
22. C. L. Holloway et al., Comparison of Electromagnetic Absorber Used in Anechoic and Semi-anechoic Chambers for Emissions and Immunity Testing of Digital Devices, *IEEE Transactions on Electromagnetic Compatibility,* Vol. 396, No. 1, pp. 33–47, February 1997.
23. IEEE 1128:1998—IEEE recommended practice for radio-frequency (RF) absorber evaluation in the range of 30 MHz to 5 GHz.

CHAPTER 4

THE CHAMBER ENCLOSURE

4.1 INTRODUCTION

Shielding serves two basic functions: first, preventing interference and second, preventing electronic eavesdropping. The type of shielding required is a function of the purpose or use of the equipment to be shielded. High-performance shielding is required where sensitive equipment must be protected from nearby high-power radar, but only moderate shielding may be required to control the electromagnetic environment within an anechoic test chamber.*

The shielding of anechoic test facilities can take a variety of forms. The type of shielding required is a function of the nature of the test requirements. If it is primarily used to establish a reflective backing for the anechoic material, then 28-gauge galvanized steel is generally quite adequate. If it is to provide a TEMPEST† environment for the test equipment of the facility, then a more elaborate shielding system may be required. If it is to test high-power equipment or the electromagnetic compatibility of an aircraft, then a welded enclosure is probably required. The remainder of this chapter will provide a brief review of the types of enclosures commonly used to house anechoic test facilities. There are limitations with each type, depending on the type of performance required and the length of service expected. The references provide a detailed review of the various construction methods.

*A complete treatment on shielded enclosures is available in the book, *Architectural Electromagnetic Shielding Handbook*. Further information is available in Refs. 2 and 3.

†TEMPEST is a term related to the control of emissions from computer equipment processing classified information.

4.2 ELECTROMAGNETIC INTERFERENCE

Proper control of conductive and radiated interference can reduce the ambient noise level that can interfere with the measurements being conducted in the chamber.

4.3 CONTROLLING THE ENVIRONMENT

A variety of shielded enclosures can be used to control the testing environment. The three most common are the sheet metal lined, prefabricated, and welded enclosures. Shielding materials commonly used in the construction of shielded enclosures consist of the following:

1. Plywood/particleboard panels laminated on one or both sides with various grades of galvanized sheet metal are commonly used for prefabricated enclosures. Other metals, such as copper foil or screen, are also fabricated in a similar manner.
2. Various thicknesses of sheet metal are welded into place on a steel supporting structure.
3. Aluminum foil is installed with contact adhesive, or moisture-proof sheetrock, which has aluminum foil bonded to one face of the board, is used.
4. Copper foil with a paper backing which is installed with contact adhesive or a nonwoven copper material installed in like manner can form a single shield system.
5. Copper screening is mounted to wooden studs and spot soldered.
6. Galvanized sheet metal is mounted to plywood walls, resulting in a single-shield system.
7. A combination of the above is used for specialized applications.

Generally, the basic shielding effectiveness of all the shielding material is adequate for most applications except in the case of high power requirements or TEMPEST. Special techniques must be applied to meet these requirements.

The seams and doors of a shielded enclosure generally set the performance of a given enclosure.

4.4 ELECTROMAGNETIC SHIELDING

4.4.1 Introduction

As noted previously, the various forms of electromagnetic shielding generally fall into three classes, welded, prefabricated, or architectural. The latter form consists of shielding that is built into the parent structure.

4.4.2 The Welded Shield

The welded seam is the most reliable. Various welded seams are illustrated in Figure 4.1. It is also the most expensive. The steel must have a minimum thickness, usually 16 gauge or thicker, and must be field-welded. A continuous metal inert gas (MIG) weld must be made along every seam. The welds must be watertight. No pinholes are permitted. With the aid of seam leak detectors developed for testing seams, RF-tight welds are routinely achieved by the shielding industry. Well-constructed welded enclosures can provide over 120 dB of shielding effectiveness throughout the entire frequency spectrum. Shielded enclosures are buildings within a building, and the shielding is not dependent on the type of construction used for the exterior of the facility.

The construction of welded enclosures is not a new technology. Many facilities have been built to counter electromagnetic threats or control emissions. The keys to a successful enclosure are to design for constructability, specifications with unambiguous performance requirements, good-quality control provisions, well-qualified welders, and properly trained quality control inspectors [4]. Areas requiring particularly close attention include the following.

- *Floor Shield Design.* Buckling of the floor shield due to heat applied during the seam welding process has been the single greatest construction difficulty in many projects. Innovative approaches to avoid this problem are discussed in [4].
- *Corner Seams.* The designer should carefully detail all corner seams, particularly where three shield surfaces join. The designer must ensure that access is available to complete welds at these locations.
- *In-Process Weld Testing.* An active in-process weld-testing program is absolutely necessary during shield assembly to avoid systematic procedural mistakes, which could lead to costly repairs later.

A variety of welding techniques are used in shield construction, and each should be reviewed prior to designing a welded installation. A complete discussion of the various techniques is available in references [1, 4].

4.4.3 The Clamped Seam or Prefabricated Shield

The most common shield seam is the clamped seam. Of these, most manufacturers of shielded enclosures use the geometry illustrated in Figure 4.2. The design is called "hats and flats." The part that is used to hold nuts or is screwed into is in the form of a hat channel, whereas the interior strapping is flat, with hole spacing for the heavy screws used in these installations. Where the shielding is mounted on both sides of the plywood or particleboard panels, the shielding effectiveness can reach over 100 dB through 18 GHz. This system is capable of operation up to 40 GHz, with proper attention to doors and vents. A completed installation is illustrated in Figure 4.3. Complete details on this form of shielding are detailed in Ref. 1.

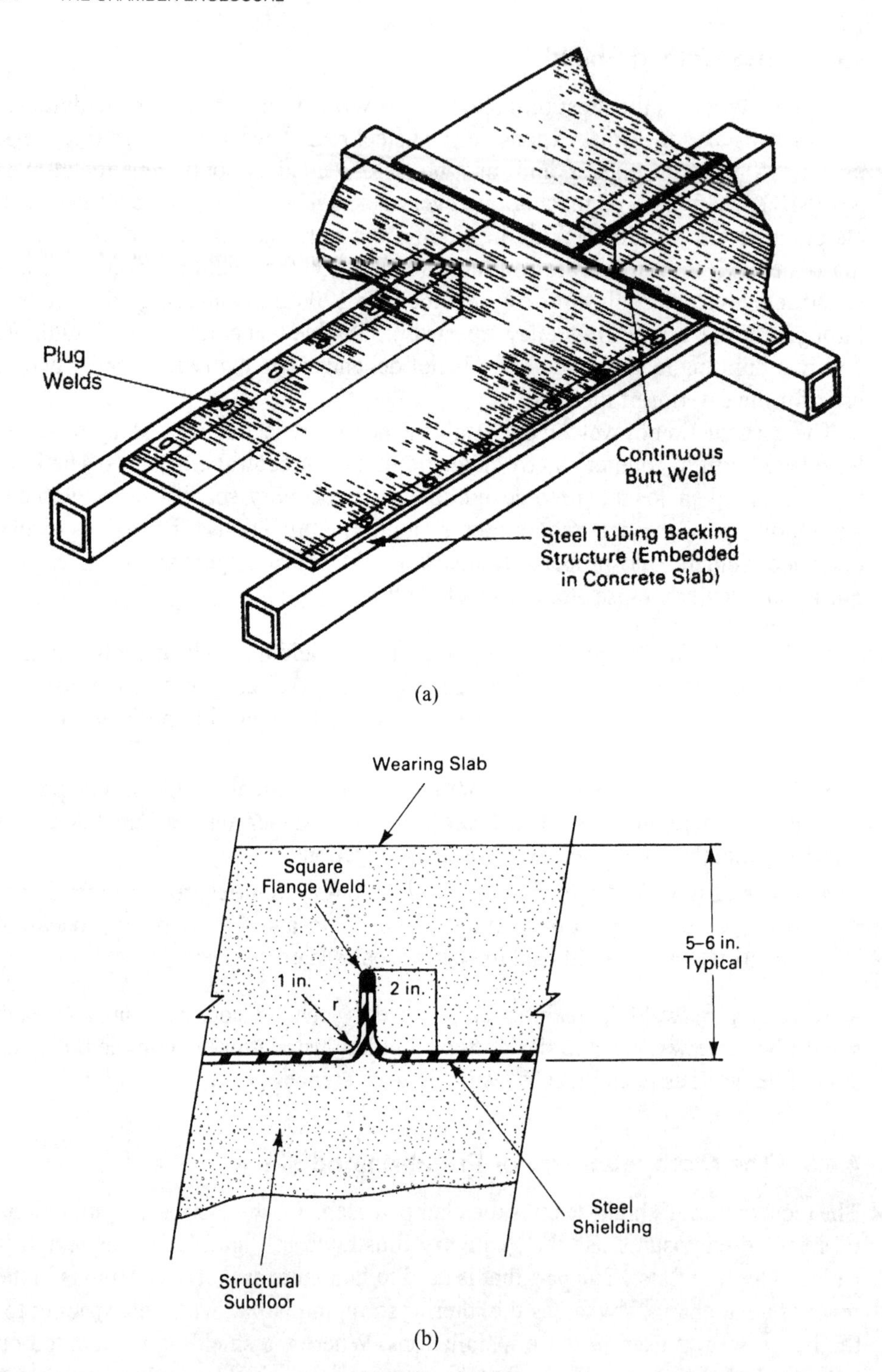

Figure 4.1 Selection of welded seams. (a) Method of plug welds and backup structure, used in floors, ceiling, and walls. (b) A method of using pan-like panels to form the floor in a shielded enclosure.

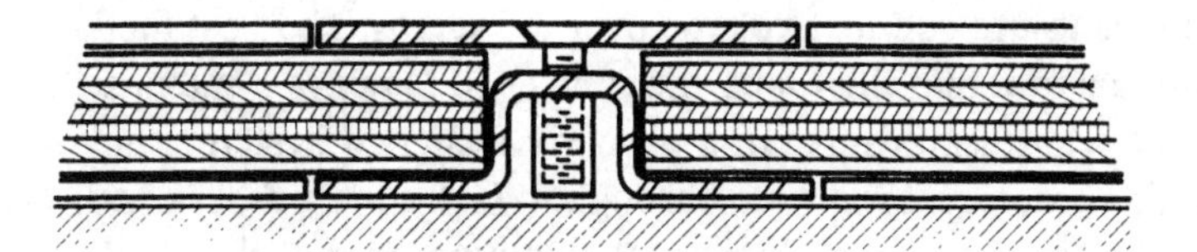

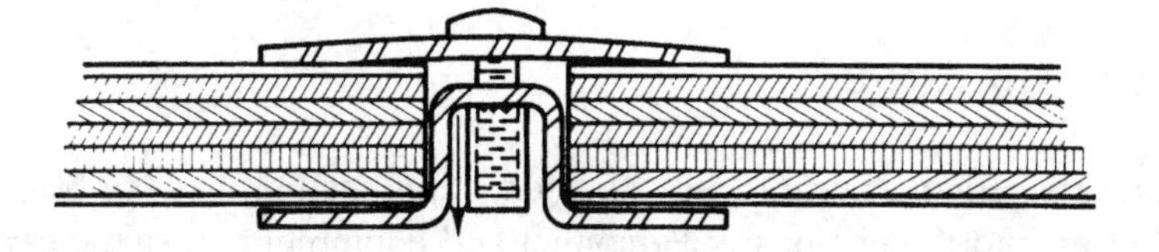

Figure 4.2 The clamped seam.

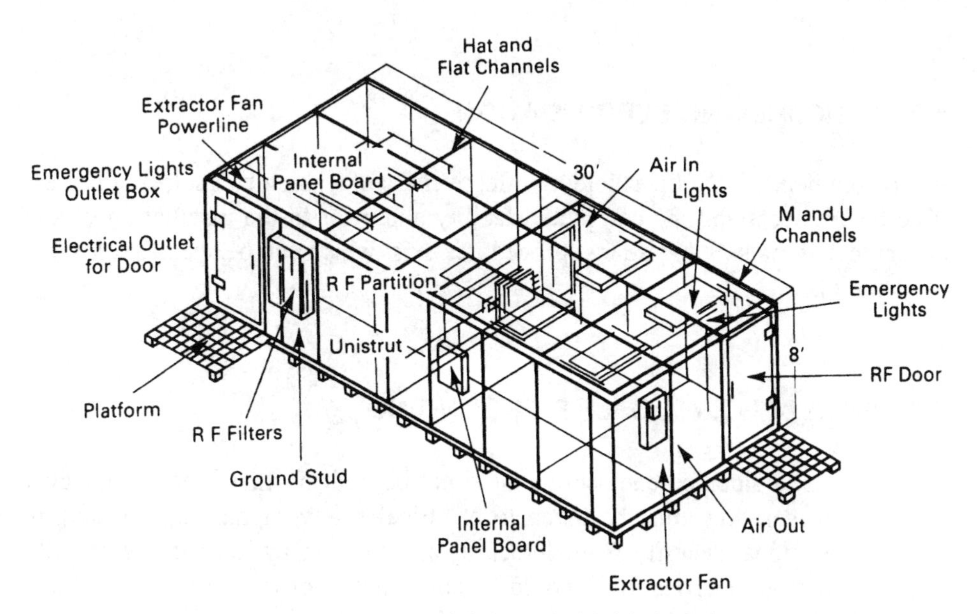

Figure 4.3 The prefabricated shielded enclosure.

4.4.4 The Single-Shield Systems

There are a variety of methods used to accomplish single-shield systems. The most common is to use conductive tape over the seams. The best method is to lap the shielding material and then tape the joints. The goal is a metal-to-metal seal throughout the whole enclosure. In a properly designed system, shielding effectiveness on the order of 60 dB up through 1 GHz can be obtained. Generally speaking, this form is called architectural shielding, in that the shielding is built into the structure of the existing building. A large variety of techniques have been developed to accomplish this type of shielding [1].

4.5 PENETRATIONS

Finally, a shielded enclosure is only as good as the method used to bring the various services into the enclosure. Personnel and equipment entry, power, ventilation and heating, high-pressure air, fire sprinklers, data and control lines each require specialized means of penetrating a shielded enclosure. They must be properly designed and implemented to ensure that proper shielding effectiveness is achieved in the final installation. The references provide guidance on the design of such penetrations. The shielded door is the most critical of all the penetrations. Various forms of the door are illustrated in Figure 4.4. These are precision electromagnetic–mechanical devices that must be designed, installed, and maintained with extreme care. Specials designs maybe required for anechoic chambers, especially those that use ferrite, because its use may double the weight of the door.

4.6 PERFORMANCE VERIFICATION

It is recommended that the shielding effectiveness of a shielded enclosure be verified before accepting the completed facility and prior to the installation of any anechoic materials. Standard test procedures exist which can be specified for this purpose [5].

4.7 SHIELDED ENCLOSURE GROUNDING

Grounding of a shielded enclosure should not be a hit-or-miss kind of endeavor [6, 7]. Consideration must be given to electrical safety, signal control, and, in some cases, signal security. Signal security is governed by the various security agencies, whose specifications should be consulted prior to construction of any type of facility that will be processing classified information.

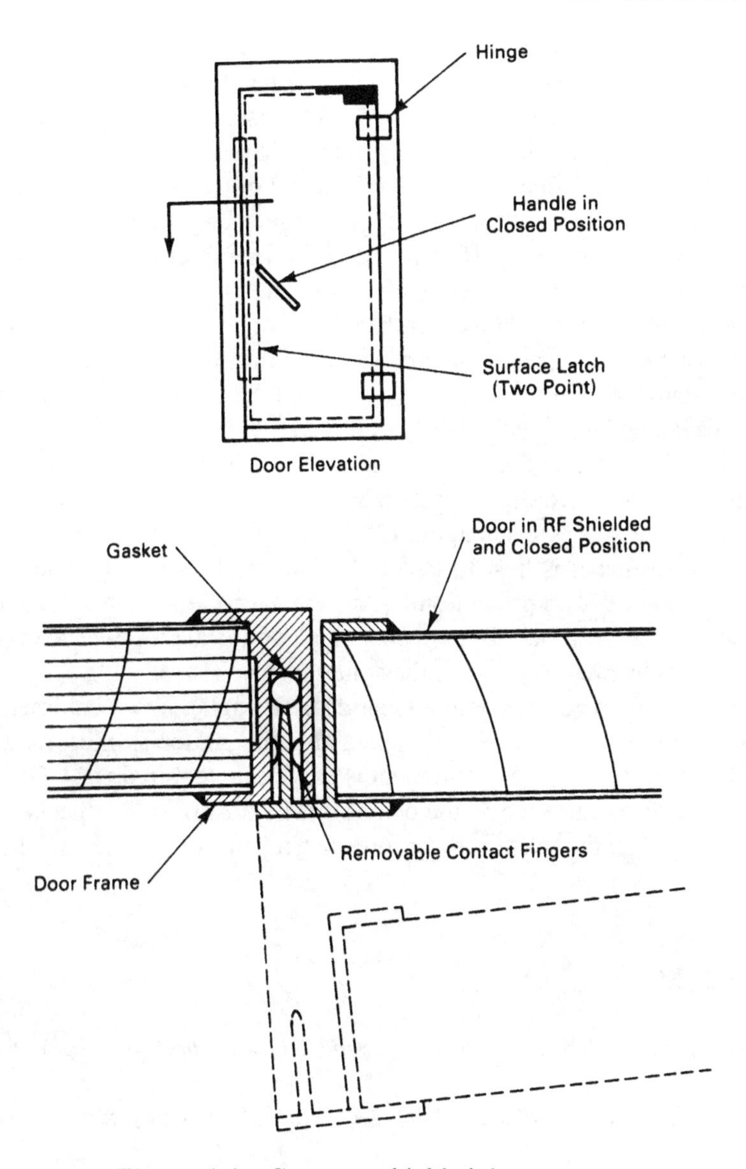

Figure 4.4 Common shielded door systems.

4.8 FIRE PROTECTION

The fire protection requirements are specified by the insurer of the building in which the chamber is to be located and by the local fire department. Three versions are generally used. (1) *Sprinkler systems:* In general, it is desirable that sprinkler heads not be present inside an anechoic chamber because they become sources of reflected energy especially in radar cross-section test facilities. One common version is the telescopic system that has special sprinkler installations that are designed to pop down when charged with water upon the activation of a smoke detector within the chamber. These are generally set up to be "dry" piped systems. Unfortunately, the history of these systems has been poor, because accidental discharge of these type of water systems have caused more damage than fires, few of which have occurred since the advent of the NRL 8093 flammability requirements in 1977. The absorbers act like large sponges and are very difficult to dry out if soaked by water. If standard sprinkler systems must be used, then care should be taken to minimize the effect of reflections from the standpipes by locating them off the chamber axis in the ceiling of the chamber. Recently, some suppliers have been using phenolic pipe in lieu of the normal black pipe, especially in EMC chambers. (2) *Gas discharge systems:* A number of Halon substitutes have been developed which smother fires but do not damage the equipment in the chamber. CO_2 has been used in vehicle test chambers, but care must be taken in the design of the fire protection system so that personnel are not caught in the chamber when the gas is discharged. (3) *Containment:* Some facility managers have determined that it is best to isolate the facility behind two-hour fire-rated walls and not use any fire protection equipment except elaborate smoke detector systems within the chamber. Some recent shielded installations have used sprinklers above the chamber to enhance the containment.

Because chamber fires generate a great deal of smoke, it is recommended that smoke inhalation equipment be maintained at a fire station immediately outside the chamber access door. This equipment is especially required if a CO_2 fire control system has been installed in the chamber. The area must be checked prior to manually activating the system to ensure that no personnel are still inside the chamber.

REFERENCES

1. L. H. Hemming, *Architectural Electromagnetic Shielding Handbook,* IEEE Press, New York, 1992.
2. R. Morrison, *Grounding and Shielding Techniques,* John Wiley & Sons, New York, 1998.
3. L. T. Genecco, *The Design of Shielded Enclosures,* Butterworth-Heinemann, Woburn, MA, 2000.
4. *USAF Handbook for the Design and Construction of HEMP/TEMPEST Shielded Facilities,* AF Regional Civil Engineer Central Region, Dallas, TX, 1987.
5. IEEE Std 299-1991, *IEEE Standard Method for Measuring the Effectiveness of Electromagnetic Shielding Enclosures,* IEEE Press, New York, 1991.
6. R. P. O'Riley, *Electrical Grounding: Bringing Grounding Back to Earth,* 6th edition, Delmar Publishing, Albany, NY, 2001.
7. IEEE Std 1100-1999, *IEEE Recommended Practice for Powering and Grounding Electronic Equipment,* IEEE Press, New York, 1999.

CHAPTER 5

ANECHOIC CHAMBER DESIGN TECHNIQUES

5.1 INTRODUCTION

The design of microwave anechoic chambers (arrangement of absorber materials on the inside surfaces of a chamber enclosure) began largely empirically in the 1950s. As the concepts became more practical, commercially developed high-performance anechoic absorbers became available in the 1960s, along with enhanced performance. It was then possible to design chambers using geometrical optics (GO) techniques, also known as ray tracing. This has led to a variety of standard chamber designs that are commonly available from about a dozen chamber manufacturers throughout the world. Since the 1980s, a new class of chamber has evolved. The demand for indoor measurements at low frequencies (30–1000 MHz) has led to ferrite/hybrid lined chambers. That, in turn, required more elaborate design techniques in order to optimize the performance.

5.2 PRACTICAL DESIGN PROCEDURES

5.2.1 Introduction

The following factors need to be considered in the design of an anechoic chamber absorber layout:

a. Type of measurements to be performed
b. Frequency of operation

c. Floor space available
d. Height of the space available
e. Geometry of the chamber
f. Method of acceptance testing
g. Cost

5.2.2 Quick Estimate of Chamber Performance

To demonstrate the most common method of designing a microwave anechoic chamber, assume a general-purpose anechoic test facility for measuring small aperture antennas. Typical antennas would be small horns, spirals, log-periodic dipole arrays (LPDA), and sinuous antennas over the 2- to 18-GHz frequency band.

Further assume that the chamber will be 3.0 m (10 ft) wide, 3 m (10 ft) high, and 6 m (20 ft) long. The range length will be on the order of 4.9 m (16 ft) as indicated in Figure 5.1. The free-space VSWR procedure will be used to perform the chamber acceptance test. (Consult Chapter 9 for complete details of the Free-Space VSWR procedure.) Figure 5.1 depicts various terms involved in a simple ray-tracing design. The design method is related to the procedures used to test the chamber. When a chamber is tested, a field probe is first run transversely across the chamber and through the center of the test region as shown in Figure 5.1 as line T. In order for the probe to detect a source of extraneous energy, it must change its phase with respect to the direct path. In the case of the transverse probe, only energy reflected off the sidewalls will be detected because all other surfaces are parallel with the movement of the probe. Thus we need only consider the energy reflected from these surfaces to determine the amount of reflected energy that reaches the test region via these surfaces. The other item that must be considered is the directivity of the source antenna, because the level of illumina-

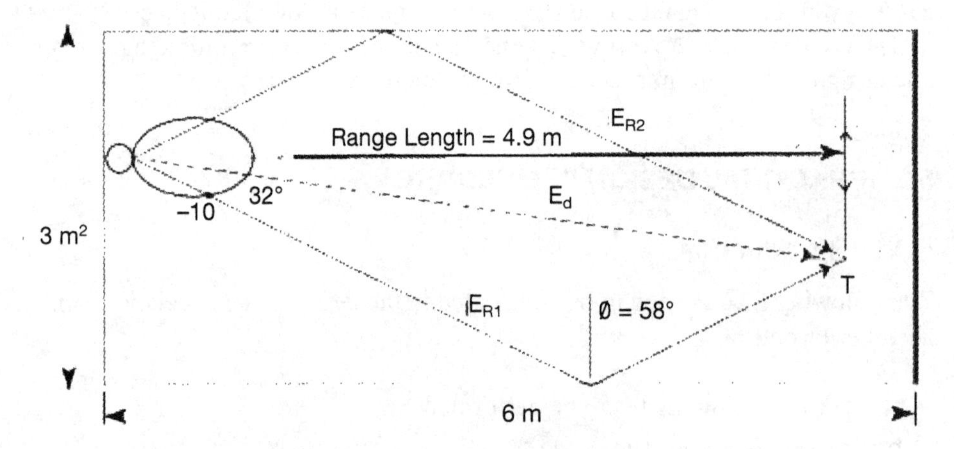

Figure 5.1 An illustration of a simple ray-tracing design.

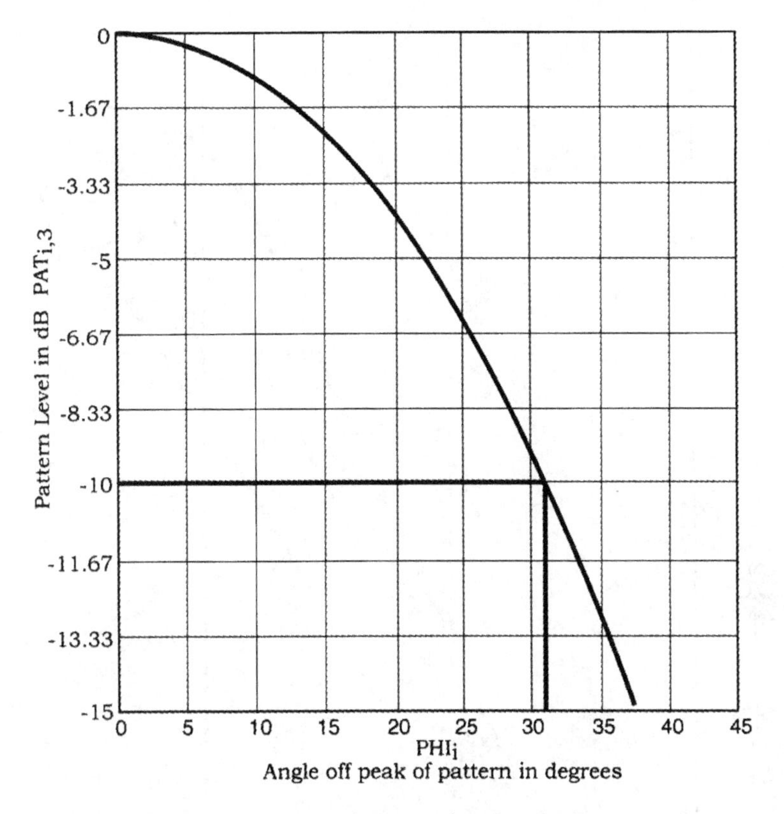

Figure 5.2 Chart of pattern levels of a horn antenna.

tion on the sidewalls is a function of the angle off the main beam of the source antenna. For the geometry assumed, this would mean that the pattern level at 32 degrees off the main beam would be the level illuminating the sidewall surfaces about the point of reflection. If one assumes that the antenna being used is a source horn antenna with a 28 degree half-power beamwidth, then the pattern level would be on the order of –10 dB,* as indicated in Figure 5.2. The angle of incidence at the sidewall absorber is 58 (90 – 32) degrees. Assume that we require the extraneous signal level to be a minimum of –40 dB at 2 GHz. Then the sidewall absorber must be 40 – 10 or 30 dB at a wavelength of 15 cm (5.9 inches). From Figure 5.3 the absorber thickness for 30-dB reflectivity at 58 degrees is found to be 2.8 wavelengths thick or 45.7 cm (18 in.) for 2 GHz. Note that we did not consider the reflected signal level in the test region to be the sum of two rays for the transverse test geometry. The reason is that the standard approach in rectangular chamber design is to construct the specular regions of absorber on the sidewalls with the absorber arranged on the diagonal. That is, the absorber blocks are rotat-

*When the pattern level is found to be greater than –10 dB, this level is assumed to allow for sidelobe levels not evident in the calculated pattern.

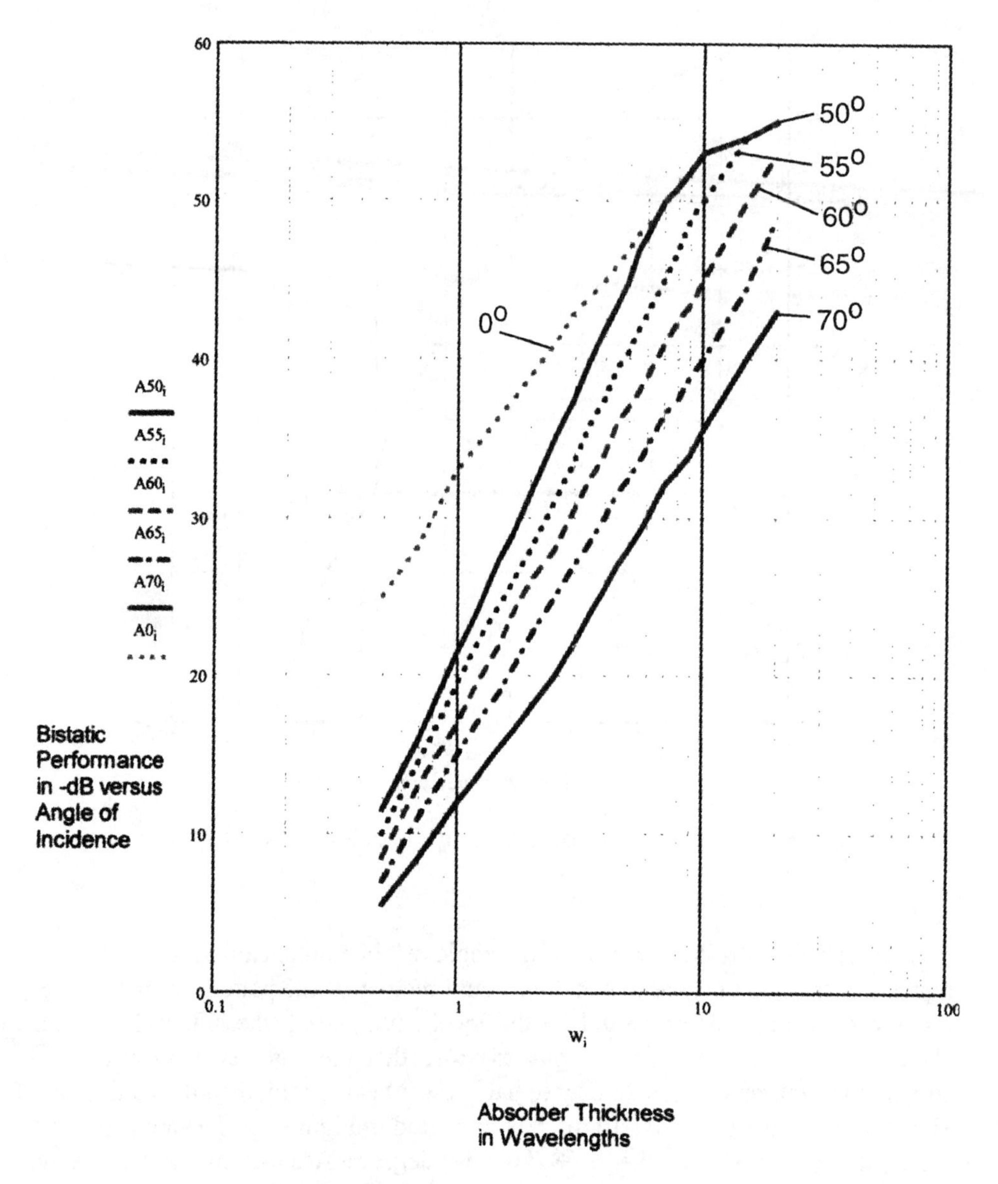

Figure 5.3 Wide-angle performance of pyramidal absorber.

ed 45 degrees and mounted in a long patch on the chamber's reflecting surfaces. This rotation drops the forward scatter by approximately 3–6 dB (i.e., approximately the amount added by the reflections from the second wall) per wall surface with respect to the reflection levels given in the curves of Figure 5.3. The end result permits consideration of a single ray reflected off the sidewall as adequate to estimate the performance. The effect of the source antenna pattern on the sidewall illumination is the second factor required for estimating the amount of absorber performance needed for a given reflected signal level in the test region. The required backwall absorber value is simply the reflectivity of the normal incidence performance, and its value can be taken from Figure 5.4. For 40 dB at 2 GHz, this requires a thickness of three wavelengths or a thickness of approximately 45.7 cm (18 in.). The actual design is a function of the total chamber geometry and is discussed in detail in the following chapters for each type of chamber. This simple approach works adequately for microwave chambers where the rooms are large in terms of a wavelength. But when the wavelength is long and hybrid absorbers are used in the designs, more elaborate analysis is required.

5.2.3 Detailed Ray-Tracing Design Procedure

Assume that a free-space continuous-wave (CW) RCS chamber is required. Further assume that the size of the room has been determined by the size of the test items and the frequency range. The following procedure can be used to estimate the worst-case reflected signal level in the test region [1].

The source antenna and the target are on the longitudinal axis of the chamber. Because the absorber on the chamber wall is not perfectly absorbing, energy can propagate from the source antenna to the test region by many paths. Several of these are illustrated in Figure 5.5; in this figure, the more important rays reaching

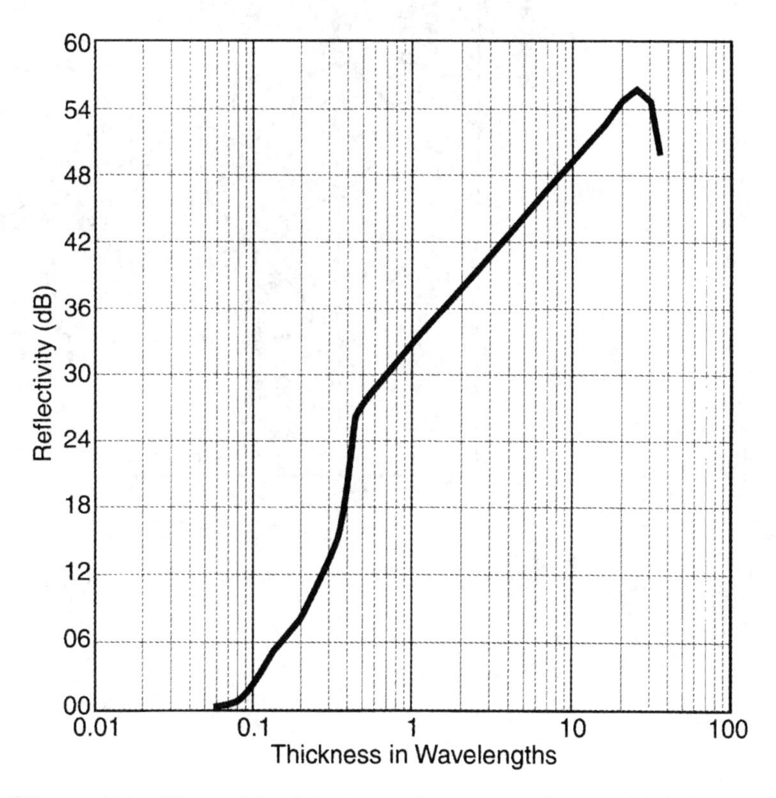

Figure 5.4 Normal incidence performance of pyramidal absorber.

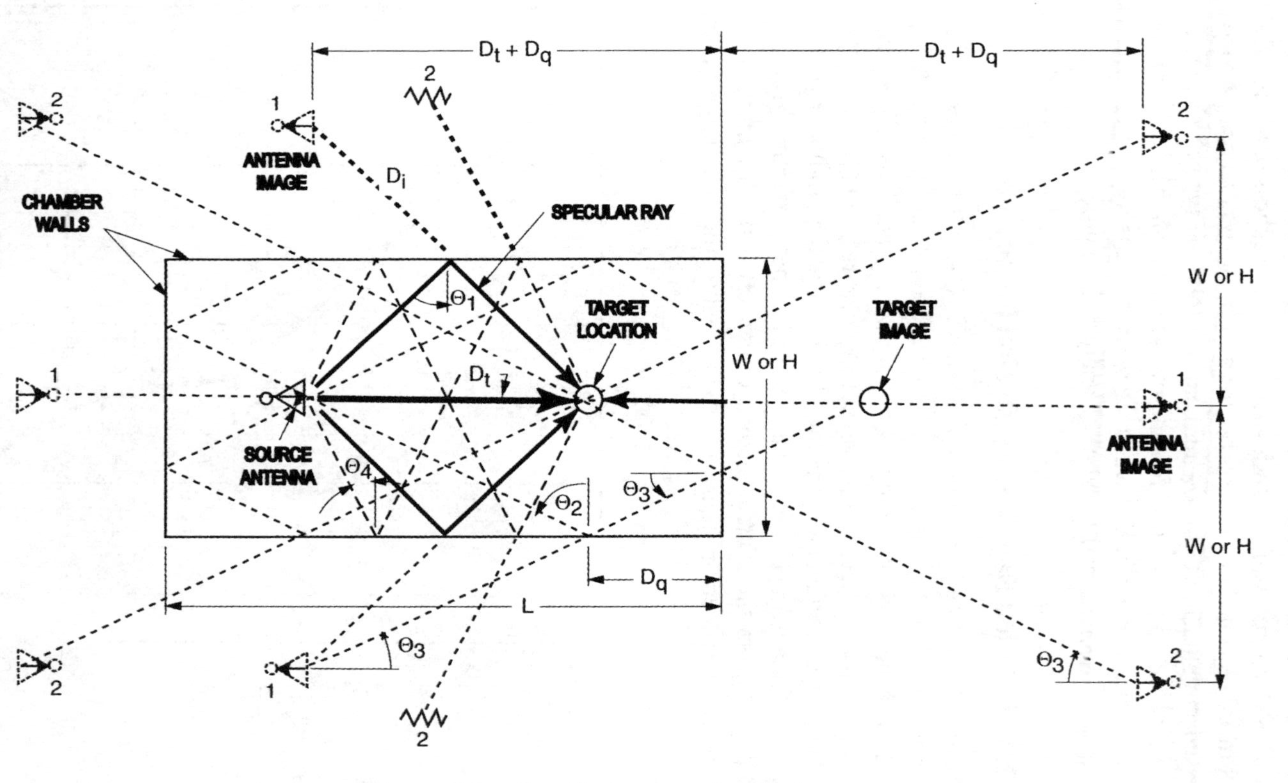

Figure 5.5 Detailed ray-tracing design.

the test region are drawn as heavy solid lines. The first is the direct path ray from the source to the test region. Of next importance are the single-reflection specular rays from the side walls (including the ceiling and floor). The rays are drawn all the way through the absorber to the chamber walls. This is the conventional method to define the incidence angle when the absorber manufacturer measures the absorber reflection coefficient. The fact that the actual rays may be refracted at the absorber surface is ignored. (This would make the actual rays slightly longer than drawn in the figure.) Also ignored are any phase shifts caused by the propagation velocity in the absorber, which is slower than in free space. Next, is the single-reflection, specular ray from the end wall directly illuminated by the antenna main lobe. The wall behind the source antenna is next considered. Finally, the two-bounce rays hitting both the side and end wall are factored in.

Each of the reflected rays can be analyzed as having come from an image of the source antenna. These images are drawn with dotted lines in Figure 5.5. The number by each image antenna indicates the number of specular reflections associated with it. The images associated with single-bounce reflections are simply the antenna's mirror image in the wall that the ray hits. The images associated with the two-bounce reflections are formed by first taking the antenna image in a side wall and then finding the image of this image in an extension of the one end wall (drawn with a dashed line in Figure 5.5). This process can be continued ad infinitum to account for other multiple bounces. Those images and associated reflected rays are of smaller magnitude than the ones that are shown, because the ray is further attenuated at each reflection point.

Figure 5.5 illustrates one advantage of the image concept. This is the ease with which the rays can be drawn. The line joining each first-order (single-bounce) image with the real antenna is perpendicular to the respective wall, and the real antenna and its image are equidistant from the wall. Similarly, first-order images are imaged in other walls (or extensions of other walls) to form the second-order (double-bounce) images. A straight line drawn from each image to the target location defines the direction of incidence on the target and defines the point of specular reflection at the wall. This provides a simple geometry for calculating incidence angle on the absorber. For example, the double-bounce image at the lower right in Figure 5.5 permits rapid calculation of the incidence angle θ_3 as $\arctan(W/(D_t + 2Dq))$.

However, we can use the line drawn from the first-order image at the figure bottom to the target image to arrive at the same answer. The part of this latter line that lies within the chamber defines one leg of the actual ray. The incidence angle labeled θ_2 is the complement of θ_3.

The second advantage of the image concept is that, once the locations of the antenna images are determined, simple free-space propagation equations can be used to calculate the field contributions from each image antenna at the target. That is, it is assumed that the chamber walls do not exist. The target is illuminated by a group of image antennas as well as the source antenna. Each image antenna is assumed to be excited, with less power than the source antenna by the amount of the wall losses. The directivity patterns of the image antennas are also taken into account, explained below.

The single-bounce rays could be easily constructed in Figure 5.5 without resorting to the use of antenna images. If either the source antenna or the target, or both, are off the chamber axis, then the image concept becomes much more useful for drawing even the single-bounce rays.

If the chamber walls were to reflect perfectly, then each image antenna would be excited with the same signal strength as the source antenna. However, the image antennas are treated as being less strongly excited than the source antenna by a factor just equal to the absorber reflection coefficient at the point when the ray reflects from the wall. Figure 5.3 and Figure 5.4 can be used to determine how many decibels each image is down from the source antenna. For the two-bounce images, the absorber reflection coefficients are found at each reflection point on the wall, and the two separate decibel values are added together to find the total image strength.

An additional consideration is the effect of the source antenna directivity on the reflected rays. The source antenna is aligned so that the peak of its beam is directed toward the center of the test region. The antenna pattern illuminates the walls with weaker signal amplitudes. The image antennas have the same pattern as the source antenna, and the angle from the image boresight axis to the respective ray is the same for each image antenna as for the source antenna. The antenna gain, in decibels relative to the pattern peak, is algebraically added to the wall reflectivity discussed previously. The antenna pattern can be determined from measured or calculated antenna patterns. For an example, refer to Figure 5.2. A final factor to be considered is the differential path length between the direct path ray and the reflected paths. The propagation factor to be applied to each reflected ray is

$$P = 20 \log(D_t/D_i) \quad \text{(dB)} \tag{5-1}$$

where
D_i = image distance from the respective image to the target location
D_t = distance from the source antenna to the target location

The propagation factor (P) is added to the antenna relative gain (G) and to the wall reflection coefficient (R), to find the amplitude of each reflected ray reaching the target location. Each factor is expressed in decibels and is a negative number. Expressed as an equation, we have

$$T = R + G + P \quad \text{(dB)} \tag{5-2}$$

where
T = total strength of each multipath ray reaching the test region center in decibels relative to the direct ray.

R = reflection coefficient of the specular ray where it reflects from the absorber, in decibels relative to a metal plate. If more than one bounce off of absorber is involved, then R is the sum of the reflection coefficient at each specular reflection.
G = antenna-pattern gain in decibels relative to the peak of the beam.
P = propagation (or path length) parameter from Equation (5-1) in dB relative to the direct ray.

Assume a specific chamber as shown in Figure 5.6. The reflectivity level, as determined by the Free-Space VSWR Test Method at the center of the test region, is estimated. Figure 5.6 shows several possible paths that multipath rays can travel from the source antenna to the center of the test region for a monostatic RCS test system. Rays A, B, and C involve only one reflection off the respective wall, whereas rays D, E, and F involve reflections off two walls. Only a single ray of each type is drawn. There will be three additional rays each of ray A and rays D through F. That is, ray A is shown reflecting off one side wall, and there will be other similar rays reflecting off the other side wall, off the ceiling, and off the floor. Rays involving three or more reflections are not drawn, because they would be of negligible amplitude. Also not drawn are multiple-bounce rays that cannot be drawn in a horizontal or vertical plane. For example, a ray could hit high on a side wall about halfway from the antenna to the target, then bounce up to the ceiling and then to the target. Such double-bounce rays could be comparable in magnitude to the double-bounce rays that are drawn.

Table 5.1 summarizes the factors that enter into calculating the magnitude of each ray. The antenna directivity is taken from measured patterns of a horn anten-

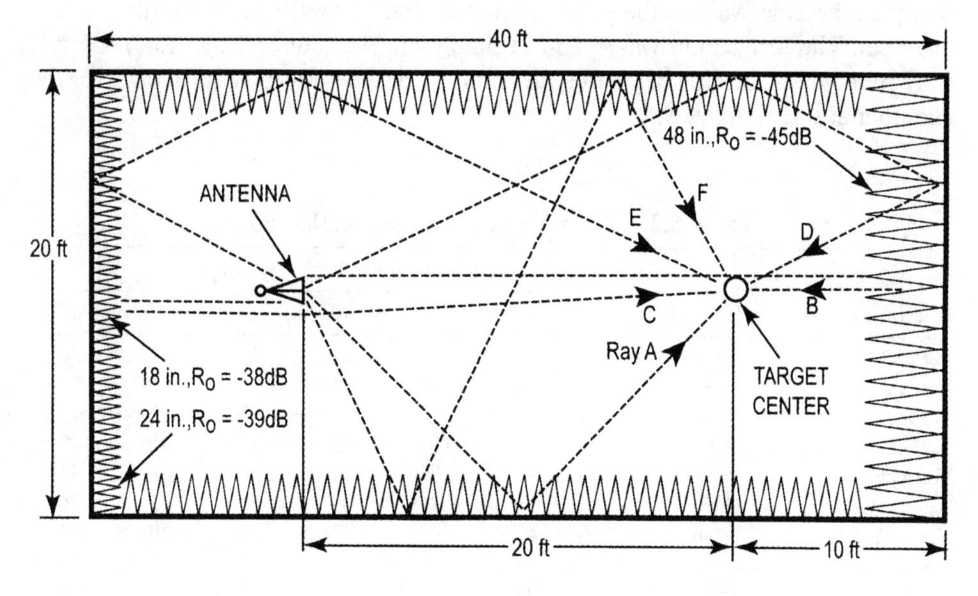

Figure 5.6 A specific anechoic chamber design.

Table 5.1 Example of Reflectivity Analysis

Ray Designation	Number of Rays	Number of Bounces	Antenna Directivity (dB)	First Reflection (dB)	Second Reflection (dB)	Path Factor (dB)	Ray Magnitude (dB)
A	4	1	–6	–32	—	–3	–41
B	1	1	0	–45	—	–6	–51
C	1	1	–26	–36	—	–6	–58
D	4	2	–2	–32	–43	–7	–75
E	4	2	–20	–34	–23	–7	–84
F	4	2	–13	–37	–37	–7	–94

na similar to those that will be used in the chamber. Also shown in the last column of Table 5.1 is the resultant value for each of the rays (normalized to the magnitude of the direct ray coming from the source antenna to the center of the test region). It is seen that all of the double-bounce rays are at least 30 dB smaller than ray A and, therefore, are relatively unimportant. If we perform a vector summation of all these rays (four of ray A, one of ray B, etc.), the worst-case sum is –28 dB relative to the direct ray illuminating the target. If we assume a more reasonable assumption that the ray phases are random, then the root-sum-squared (RSS) summation gives a more expected summation of –35 dB. Thus, it was expected that this chamber design would meet the goal of –30 dB for the test region reflectivity level at 1 GHz.

To verify the analysis, measurements were made in the chamber test region at several frequencies, using the Free-Space VSWR Test Method. A transverse antenna probe was used at five locations. Three probe angles were used to evaluate the test region; ±30 degrees, ±60 degrees, and ± 90 degrees; the latter looks directly at the side wall as the probe is moved transverse to the centerline of the chamber. The test region was 1.5 m in diameter. The reduced data for 1 GHz is shown in Table 5.2. This compares favorably with the –30-dB requirement for the chamber at this frequency.

Table 5.2 Example of Measured Reflectivity

		Probe Pointing Angle vs. Reflectivity (dB)					
Polarization	Scan Line	90	60	30	–30	–60	–90
Vertical	Center	33	34	33	34	37	36
Vertical	Front	37	42	35	37	42	57
Vertical	Rear	34	33	40	37	37	43
Vertical	Top	37	39	43	43	41	40
Vertical	Bottom	37	30	41	35	42	43
Horizontal	Center	35	41	43	36	40	34
Horizontal	Front	42	41	43	36	40	34
Horizontal	Rear	43	44	45	42	35	35
Horizontal	Top	41	41	42	44	38	43
Horizontal	Bottom	40	38	42	37	32	33

5.3 COMPUTER MODELING

5.3.1 Introduction

With the advent of the personal computer and its significant computing power, it has become practical to model or simulate the anechoic design process. Ray tracing a large number of rays is now practical, and greater attention can now be given to second-order effects in the design of chambers not previously possible. One such example will be given, with references to others. Ray tracing is sufficient where the wavelength is short with respect to the size of the chamber. Where the chamber dimensions approach a wavelength, this technique fails to properly predict a chamber's performance. A more detailed approach is required, such as the Finite-Difference Time-Domain Method. An example of the use of this technique will be demonstrated and references given to aid in the use of such a system in Section 5.3.3.

5.3.2 Ray Tracing

An excellent example of using ray tracing to evaluate the performance of an anechoic chamber for a specific application is provided in Ref. 2. The report considers an electromagnetic field simulation of a large anechoic chamber and compares the results to verification measurements. The simulation is a Geometric Optics (Ray Tracing) mathematical model of the direct path between two antennas and the chamber reflections. Due to the frequency-dependent nature of the pyramidal radar absorbing material (RAM), there are two separate absorber models. The model for the frequency range of 30–500 MHz was used to characterize the lossy specular scattering. The specular scattering was modeled as a lossy, tapered, TEM transmission line in an inhomogeneous anisotropic tensor material. The frequency range from 500 MHz to 18 GHz was characterized by dominant tip diffraction of RAM patches. The model made use of a uniform Theory of Diffraction code for a dielectric corner. The measurements and simulation were based on an azimuthal cylindrical scan. Diagnostic measurements were also performed by a cylindrical scan of a directional horn antenna to locate scattering sources in the chamber.

The computer simulation project was developed to support the U.S. Air Force Benefield Anechoic Chamber which has a 24.4-m (80 ft)-diameter turntable located on the floor and the main coordinate system positioned at the center of the turntable. The chamber's internal dimensions are 21.3 m (70 ft) high, 76.2 m (250 ft) wide, and 80.8 m (265 ft) long. A pyramidal absorber 0.46 m (18 in.) thick covers all interior surfaces except for the floor, which uses 1.22-m (48-in.) material. The test region located above the turntable was evaluated using a cylindrical scan, as illustrated in Figures 5.7 and 5.8. Good agreement was found between the simulation and the measurement for the low-frequency data when the antenna characteristics were included in the simulation. Good agreement was also found for the high-frequency data except for a slight variation from a low-frequency

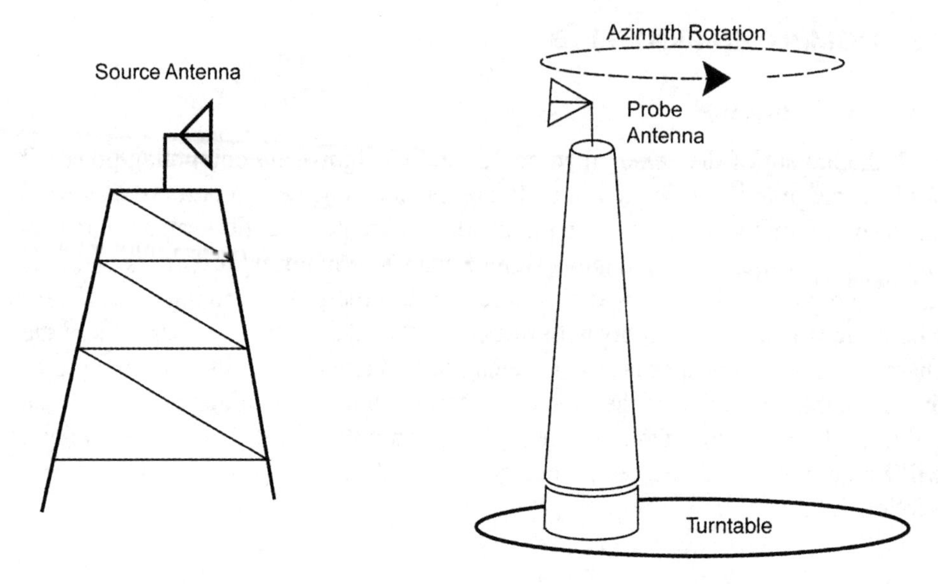

Figure 5.7 Elevation view of cylindrical scan method.

standing wave. Refer to the referenced report [2] for full details on the simulation and the measured results.

5.3.3 Finite-Difference Time-Domain Model

Where the operating wavelength is very close to the chambers interior dimensions, the ray-tracing technique breaks down and a more detailed method is required. Under a grant from the Ben Franklin Partnership Program of the Com-

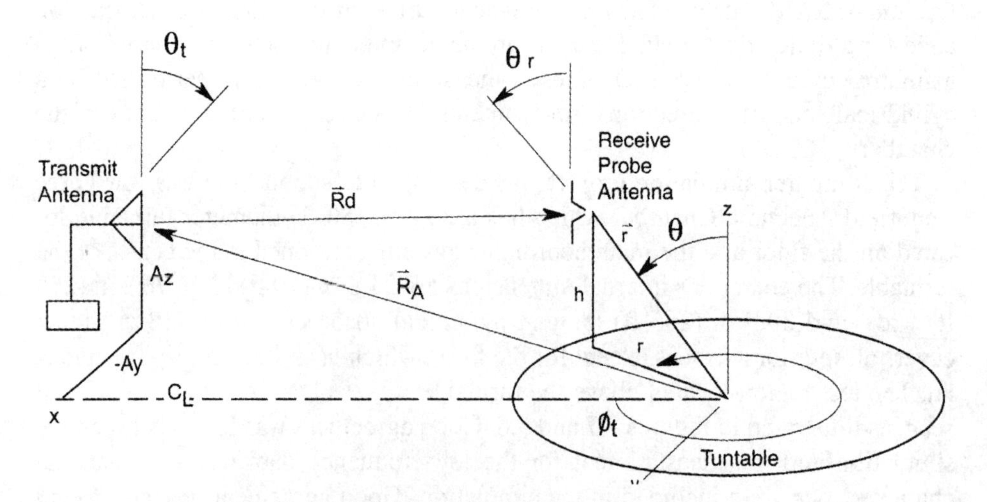

Figure 5.8 Cylindrical scan geometry.

monwealth of Pennsylvania, Lehman Chambers of Chambersburg, Pennsylvania, commissioned a set of studies. The first study [3] was with the University of Colorado for absorber models, and the second [7] was with Electro Magnetic Applications Company of Boulder, Colorado, to develop a chamber design technique for EMC test chambers using a computer model.

The first of these, referred to as the "absorber model," involves a method that allows the designer to replace a doubly periodic absorbing structure, such as urethane pyramids or ferrite grid/tile, by layering "effective" material properties. Using this method, it is possible to obtain the plane wave reflection loss of a sheet of the material [3–6]. This model runs on a personal computer and can compute the reflection coefficient at multiple frequencies in a matter of seconds. This is extremely helpful in the efficient evaluation and optimization of absorber designs. The second model, referred to as the "chamber model," uses the layered "effective" material properties in a full three-dimensional solution of Maxwell's Curl Equations by using a variant of the computer program EMA3D [7, 8]. This allows the designer to obtain the chamber performance by mathematical modeling, as opposed to trial-and-error construction on a full-sized prototype. The computational technique used is a fully self-consistent, total field solution for the electric and magnetic fields inside the chamber, as opposed to the image ray-tracing technique. The benefit of the full three-dimensional model is the ability to calculate actual performance of the chamber when testing products. The model allows for the inclusion of spatial variation and the frequency dependencies of the effective averages of the dielectric permittivities and magnetic permeabilities of different absorber materials. The model also permits modeling of transmit and receive antennas using the thin-wire capabilities available in the program. The computational volume is terminated at the metal walls of the chamber, which are treated as perfectly conducting. Using the electric and magnetic field vectors on a discretized spatial lattice within the chamber, the site attenuation and field uniformity within either a semianechoic or fully anechoic chamber can be calculated.

5.4 OTHER TECHNIQUES

A variety of anechoic chamber design techniques have been developed by various institutions. These include designs for the rectangular chamber [9–12], the tapered chamber [13] and the EMC chamber [14].

5.5 ANTENNAS USED IN ANECHOIC CHAMBERS

5.5.1 Introduction

As was pointed out in the chamber design sections, the source antenna pattern properties are a major contributor to the performance of the chamber. In the rectangular chamber, it is necessary to pick antennas that will minimize the amount of energy that reaches the side walls, but still maintain a uniform field across the test region. In the tapered chamber, it is necessary to use antennas that are compatible with operation in the conic section of the chamber. In the EMI chamber, antennas need to be calibrated very carefully, because the measurement will determine the pass or fail properties of a device under test.

5.5.2 Rectangular Chamber Antennas

In the conventional anechoic chamber used for antenna testing, the source antenna is usually located on one end of the chamber, and the item under test is located in the test region at some point down the chamber axis. The amount of extraneous energy reaching the test region is a function of the source antenna's pattern function. In a well-designed chamber the area about 30 degrees off the boresight of the antenna usually illuminates the side walls of the chamber; and that level of energy, plus the attenuation of the absorber in that area, determines the amount of energy reaching the test region. Thus, it is necessary to know the pattern properties of the antenna used as a source antenna. Generally, the antenna manufacturer specifies the antenna's half-power beamwidths over the operating band of the antenna. Both E-plane and H-plane pattern specifications are required to properly determine the side-wall illumination levels. Given the half-power beamwidth, the pattern function off the boresight axis of the antenna can be calculated by using the procedure given in Appendix B.

A variety of antennas are used as source antennas. The most common is the standard gain horn antenna. A variety of these antennas are available from various antenna manufacturers. Knowing the chamber geometry, an optimum source horn can be selected for most microwave antenna testing. The desired antenna will have a less than 0.5-dB amplitude taper across the test region. It will also have the maximum gain possible so that the pattern falls off rapidly outside the test region, thus minimizing the energy reaching the side walls, thus reducing the extraneous signal level reaching the test region. All of the standard gain horn antennas operate over the common waveguide frequency bandwidths.

An excellent source antenna is the exponential horn; it has a nominal gain of 16.5 dBi and a half-power beamwidth of 28 degrees in both planes. At 30 degrees off-axis the pattern level is –13 dB. This set of horns has the same properties across the microwave frequency range; thus it is convenient to use as a source antenna for a general-purpose antenna test chamber. Figure 5.2 shows how the pattern falls off for antennas that have similar beamwidths. If more than a general estimate of the pattern levels is required, then the specific patterns of the source antenna should be used to assess the level of illumination on the side walls of the chamber under consideration.

5.5.3 Antennas for Tapered Chambers

The tapered chamber requires an antenna that has a phase center that is electrically compatible with the conical tip used to house the source antenna. This means that the phase center can be placed such that it is on the order of less than one wavelength from the absorber surface of the cone. This requirement is discussed in Chapter 8. Horn antennas generally meet this requirement. Some low-frequency antennas, such as Yagi antennas, meet this requirement. The log periodic dipole array (LPDA) at the high end of its frequency band generally does not meet this requirement, because the antenna is smaller at the forward end of the antenna than at the back (i.e., the antenna tapered geometry is reversed to that of the tapered chamber). Ridged horn antennas make good source antennas for driving the tapered chamber at low frequencies. Some models of these antennas go down as low as 200 MHz. A good source for these antennas is companies that specialize in antennas for EMI measurements. A couple of sources exist for quadridged horn antennas. These antennas make it possible to perform both E-plane and H-plane antenna measurements with a single broadband antenna. When the source antenna is properly located in the throat of the chamber, then the geometry and absorber properties of the taper determine the amplitude taper in the test region of the chamber.

5.5.4 EMI Chambers

Antennas used in EMI chambers only have a second-order impact on the design of the chamber. A large variety of antennas are used. Dipoles, biconicals, LPAs, bi-logs, and horn antennas are all used, depending on the frequency of measurement. When radiated emissions or susceptibility are being measured, then the antennas need to be carefully calibrated. The best calibration technique is the three-antenna method [15].

REFERENCES

1. L. A. Robinson, *Design of Anechoic Chambers for Antenna and Radar-Cross-Section Measurements,* AD-B074-391L, SRI International, November 1982.
2. R. M. Taylor, *Anechoic Chamber Simulation and Verification,* Master Thesis, June 1993, California State University, Northridge, ADA308236.
3. E. F. Kuester and C. L. Holloway, Improved Low-Frequency Performance of Pyramid-Cone Absorbers for Application in Semi-Anechoic Chambers, *IEEE Symposium on Electromagnetic Compatibility,* pp. 394–399, 1988.
4. E. F. Kuester and C. L. Holloway, Comparison of Approximations for Effective Parameters of Artificial Dielectrics, *IEEE Transactions on Microwave Theory and Techniques,* Vol. 38, pp. 1752–1755, 1990.
5. E. F. Kuester and C. L. Holloway, A Low-Frequency Model for Wedge or Pyramid Absorber Array—I: Theory, *IEEE Transactions on Electromagnetic Compatibility,* Vol. X, pp. 300–306, 1994.
6. C. L. Holloway and E. F. Kuester, A Low Frequency Model for Wedge or Pyramidal Array—II. Computed and Measured, *IEEE Transactions on Electromagnetic Compatibility,* Vol. X, pp. 307–313, 1994.

7. C. L. Holloway et al., On the Application of Computational Electromagnetic Techniques to the Design of Chambers for EMC Compliance Testing, *Compliance Engineering,* March/April 1994.
8. Electromagnetic Interaction Codes (North American Edition), EMA-93-R-033, January 1994.
9. J. Gillette, RF Anechoic Chamber Design Using Ray Tracing, *IEEE AP-S International Symposium,* p. 250, 1977.
10. S. Brumley, "A Modeling Technique for Predicting Anechoic Chamber RCS Background Levels," *Antenna Measurement Techniques Association Proceedings,* p. 133, 1987.
11. S. Mishra, Computer Aided Design of Anechoic Chambers, *Antenna Measurement Techniques Association Proceedings,* p. 30–1, 1985.
12. D. A. Ryan, *A Two Dimensional Finite Difference Time Domain Analysis of the Quiet Zone Fields of an Anechoic chamber (Compact Range),* Final Report NASA NAG1-1221, Pennsylvania State University January 1992.
13. H. King and J. Wong, Characteristics of a Tapered Anechoic Chamber, *IEEE Transactions on Antennas and Propagation,* Vol. AP-15, p. 488, 1967.
14. S. Mishra and T. Pavasek, Design of Absorber-Lined Chambers for EMC Measurements Using a GO Approach, *IEEE Transactions on Electromagnetic Compatibility,* Vol. EMC-26, No. 3, p. 111, August 1984.
15. IEEE Std 149-1979, *IEEE Standard, Test Procedures for Antennas,* IEEE Press, New York, 1979.

CHAPTER 6

THE RECTANGULAR CHAMBER

6.1 INTRODUCTION

Depending on the type of electromagnetic measurement to be conducted, rectangular anechoic chambers can take many forms. They can be small bench-type units that operate in the microwave frequency range. They also can be very large shielded structures to conduct electromagnetic compatibility measurements on bomber-sized aircraft. This chapter section is organized by the type of testing to be conducted. Included are test facilities for (1) antennas, (2) radar cross section, (3) system testing, and (4) EMC testing. One important class of rectangular chamber, the compact range, is discussed separately in Chapter 7.

6.2 ANTENNA TESTING

6.2.1 Introduction

The design of rectangular chambers for antenna measurements is determined by the following factors: (1) far-field range equation, $2D^2/\lambda$ which this sets the phase uniformity across the test aperture as described in Chapter 2, (2) the minimum and maximum operating frequencies of the antenna, and (3) the uncertainty permitted in the antenna measurements. The directivity of the source antenna and the bistatic performance of the wall absorber primarily determine the level of extraneous signal level that sets the uncertainty of the measurement.

6.2.2 Design Considerations

The parameter D is the aperture size of the item under test. Given this value and the highest operating frequency, the chamber's minimum range length (R) is determined. The next parameter is the chamber width (W), which should be $> R/2$, or slightly greater than one-half the range length. That is, the chamber aspect ratio needs to be such that the angle of incidence on the adjacent walls and ceiling is on the order of < 60 degrees. This latter requirement determines the arrival angle of the incident wavefront to the side-wall, floor, and ceiling materials. This, in turn, influences the amount of attenuation experienced by the forward-scattered signal into the test region. The test region (test volume, quiet zone) diameter can then be on the order of $W/3$. The test region is the volume over which the chamber reflectivity is specified, and should be greater in diameter than the antenna to be tested.

Having determined the basic chamber size, the next step is to determine the anechoic lining required to achieve the desired chamber reflectivity or level of reflected energy in the test region. A common requirement for microwave test chambers is to set the maximum reflectivity to be –40 dB at the lowest operating frequency. This ensures that a 20-dB sidelobe level of an antenna pattern (a common antenna requirement) repeats within roughly 1 dB, regardless of the antenna location in the test region. Another approach is to determine what uncertainty can be tolerated at a given sidelobe level and then determine what the reflectivity level needs to be to ensure the required repeatability.

The uniformity of the test region field amplitude is set by the pattern of the source antenna together with the energy scattered by the chamber surfaces.

Having determined the reflectivity requirements by analysis or edict, the absorber requirements are determined as follows:

1. The receiving wall or back wall is determined directly from the absorber performance curves, because the illuminating wave front is at normal incidence.
2. The transmit wall (the wall behind the source antenna) absorber is generally about half the thickness of the receiving wall. The front-to-back ratio of the source antenna is generally sufficient to permit the use of thinner materials than is required when the wave front is at normal incidence.
3. The side-wall, ceiling, and floor absorber requirements are found by determining the angle of illumination from the source antenna and estimating the pattern attenuation directed at the specular region of the wall. This level is subtracted from the required reflectivity, and then the side-wall absorber performance is set. This level of reflectivity is found from the design curves discussed in Chapter 3 for the absorber's wide-angle performance. Generally, the low-end frequency will determine the required absorber thickness. Once the low end is set, the performance of the remainder of the frequency range can be determined from the design curves and the pattern geometry of the source antenna.

4. The area and layout of the wall absorber for the specular region is next determined. This is the area on the chamber surfaces where the energy from the source antenna reflects into the test region. Generally, the layout is based upon experience. A more analytical method is to determine how many Fresnel zones need to be covered by the high-performance material to suppress the side-wall energy to an acceptable level. The size of the area can be determined by calculating the zones of constant phase that need to be covered to provide sufficient suppression of the reflected energy. The size of the area is directly proportional to the required reflectivity. This design approach is discussed in Appendix A. Pyramidal absorber is generally used throughout the design, since it is not polarization sensitive. The remainder of the wall absorber is normally determined to be about one size down (75%) of the critical material thickness in the specular region.
5. The amplitude taper in the test region is determined from the pattern of the source antenna and the range length. Appendix B provides the steps necessary to compute the taper across the test region.

6.2.3 Design Example

6.2.3.1 Measurement Problem. Suppose that a need exists for a general-purpose microwave anechoic chamber to measure electrically small antennas from 2 to 18 GHz. Typical antennas would be horns, spirals, log-periodic dipoles, and small patch arrays. These are apertures typically less than 0.23 m (0.75 ft) in diameter. For D of 0.23 m at 18 GHz, the range length is $R = 6.4$ m (21 ft) from the transmit wall to the center of the test region. The chamber width needed to keep the angle of incidence below 60 degrees is 4.3 m (14 ft). The test region can be up to one-third of this width, or approximately 1.4 m (4.5-ft). The overall chamber length is the range length, plus the clearance behind the test region, which is assumed to be 1.22 m (4 ft) for a total of 8.3 m (27.25 ft), rounded off to 8.5 m (28 ft). This permits the measurement of a 2-GHz antenna up to 1.22 m (4 ft) in length. This is a common size for the new cell phone base station systems. Further assume that the minimum chamber reflectivity is –40 dB at 2 GHz. The receiving wall absorber, rounded to the nearest standard size, will have to be a minimum of 0.46 m (1.5 ft) thick, as determined from Figure 6.1. Therefore, the required transmitting wall absorber would be approximately 0.20 m (0.5 ft) thick as discussed in the design assumptions given in Chapter 5. Assuming a standard gain horn as a source antenna, the pattern factor would be on the order of –10 dB at 30 degrees on a 28-degree half-power beamwidth standard gain horn antenna. The design curve for 60 degrees angle of incidence (Figure 6.2) indicates that for a wide-angle performance of –30 dB, the absorber must be a minimum of 3.0 wavelengths thick. At 2 GHz, this would be on the order of 0.46 m (18 in.).

The side-wall, floor, and ceiling absorber selected above is arranged in a dia-

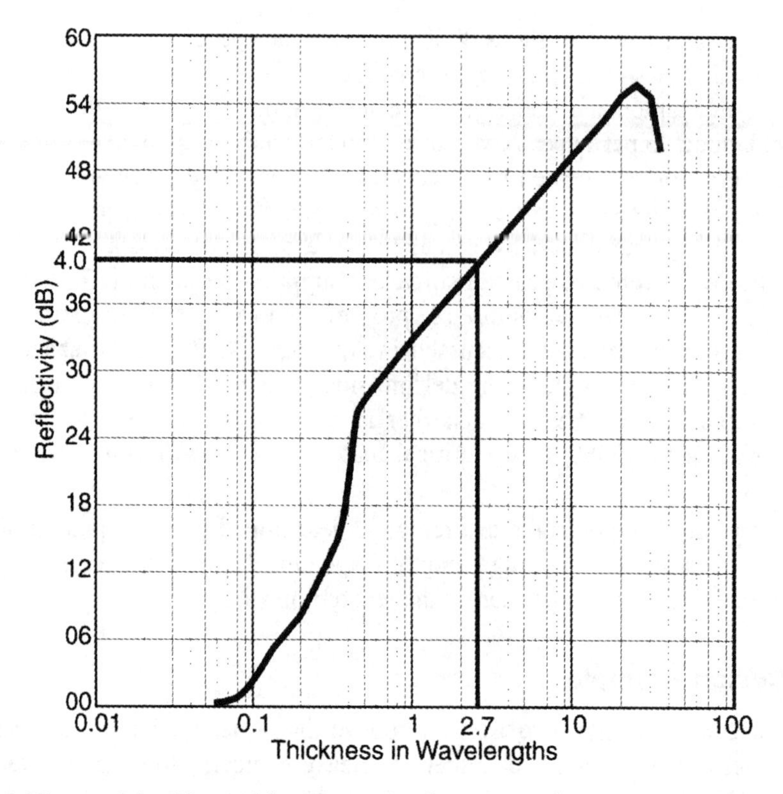

Figure 6.1 Normal incidence performance of pyramidal absorber.

mond-shaped patch by rotating the absorber 45 degrees and running the diagonal length along the chamber axis as shown in Figure 6.3. The length and width of the patch is determined using the procedures given in Appendix A. These procedures show that the specular region needs to be a minimum of 6 units long and 3 units wide and the center should be located 3.7 m (12 ft) from the transmit wall. The remainder of the absorber can be 0.30-m (1-ft)-thick units. A common method of mating in the corners is to cut the material at 45 degrees and form two-way and three-way corners as illustrated in Figure 6.4. Another method is to form blocks of material the same thickness as used in the sidewall and end wall absorbers and form a picture frame of solid material in the corners as shown in Figure 6.5. The latter is recommended for most applications, because this provides higher absorber losses in the corners and reduces potential corner reflection effects.

Practical Example No. 1. The chamber illustrated in Color Plate 1 (a separate color section is provided of selected photographs of a variety of anechoic chambers) is a rectangular chamber 5.5 m (18 ft) wide, 5.5 m (18 ft) high, and 9.1 m (30 ft) long. The design requirements for the empty chamber were as follows:

Frequency (GHz)	Reflectivity (dB)	Expected Performance
0.400	–20	–22
0.800	–28	–30
2.000	–42	–44

The source antenna at the low frequencies is a Yagi antenna. Horns are used at the higher frequencies.

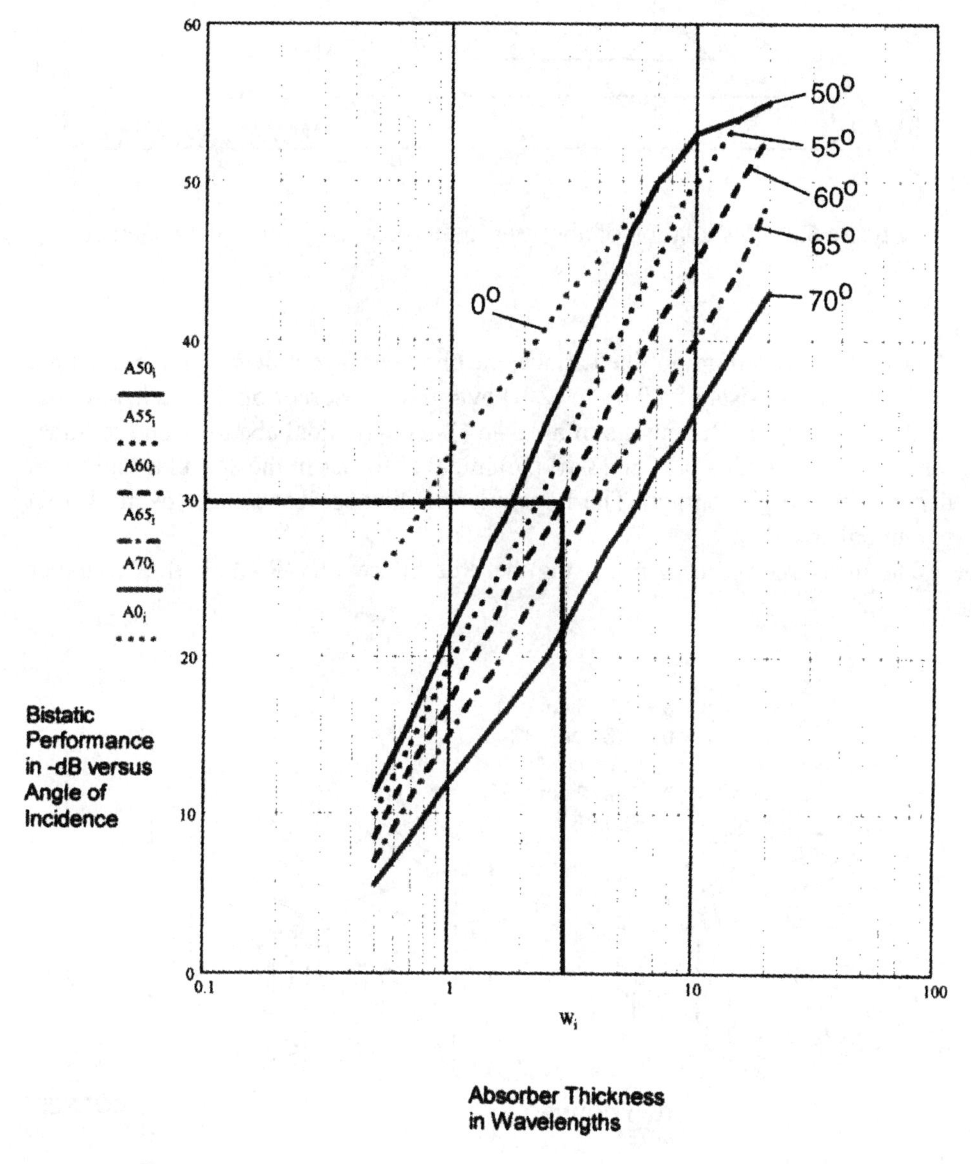

Figure 6.2 Wide-angle performance of pyramidal absorber.

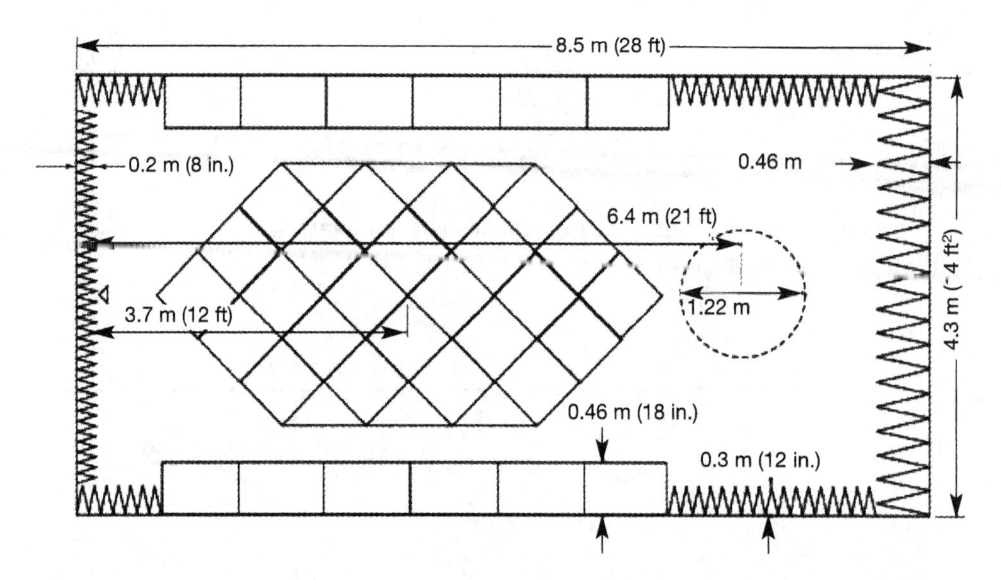

Figure 6.3 Installation of absorber in the specular region of a chamber.

Using the procedure outlined above, the chamber layout developed to meet the requirements consists of a 0.61-m (2-ft) pyramidal absorber on the back wall, behind the test region. It consists of a 0.3-m (1-ft) pyramidal absorber on the transmit wall, along with a 0.91-m (3-ft) pyramidal absorber in the specular region, as depicted in the photograph. The remainder of the absorber is a 0.46- m (1.5-ft) pyramidal absorber.

The measured performance using the Free-Space VSWR Method of chamber

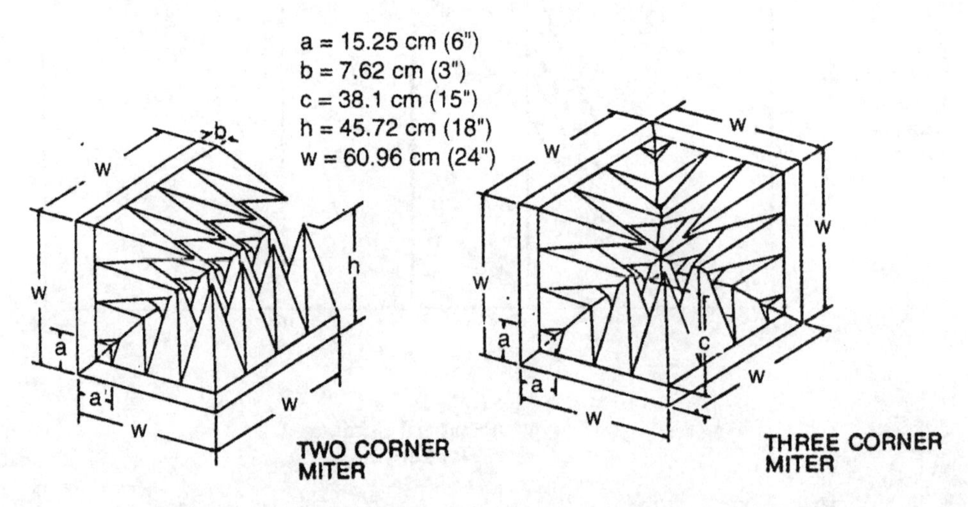

Figure 6.4 Mating of absorber in two-way and three-way corners.

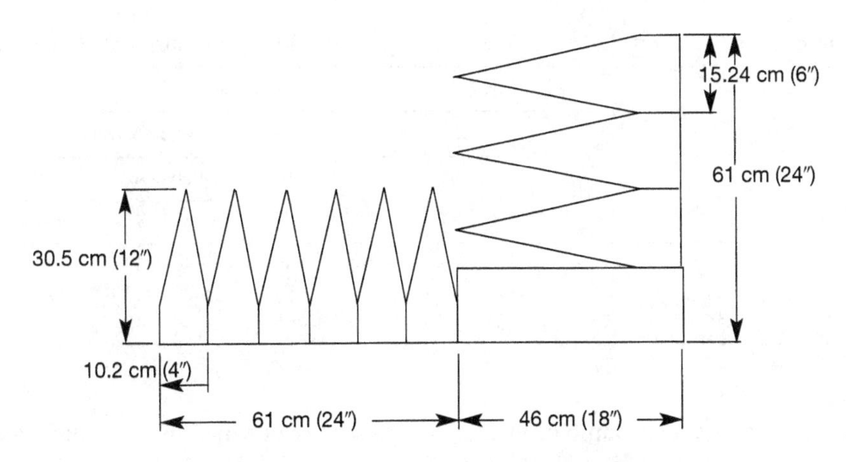

Figure 6.5 Picture frame installations.

evaluation over a 2.4-m (8-ft) test region is listed in Table 6.1. Some interaction with an antenna positioner was experienced.

Practical Example No. 2: This is a high-frequency chamber designed for the 2- to 18-GHz frequency range. Chamber is 3.7 m (12 ft) wide, 3.7 m (12 ft) high, and 7.4 m (24 ft) long. The anechoic layout is 0.46 m (1.5 ft) pyramidal on the backwall, 0.20 m (0.5 ft) pyramidal on the transmit wall, and 0.30 m (1 ft) uniform treatment on the remaining surfaces.

The specifications, expected, and measured performance of the high-frequency rectangular chamber is given in Table 6.2 over a 1.2-m (4-ft) test region:

Again, some interaction with an installed antenna test positioner was experienced during the field probing.

6.2.4 Acceptance Test Procedures

The Free-Space VSWR Test Method and the Pattern Comparison Method [1] are the two most common procedures used to establish the extraneous signal level

Table 6.1 Measured Performance of a Low-Frequency Rectangular Chamber

	Measured							
	Typical				Worst Case			
	Transverse		Longitudinal		Transverse		Longitudinal	
Frequency (GHz)	H	V	H	V	H	V	H	V
0.400	−22	−26	−40	−34	−19	−23	−33	−33
0.800	−31	−39	−47	−53	−27	−36	−43	−48
2.000	−45	−45	−49	−52	−41	−43	−43	−48

Table 6.2 Measured Performance of a High-Frequency Rectangular Chamber

Frequency (GHz)	Specified dB	Expected dB	Measured							
			Typical				Worst Case			
			Transverse		Longitudinal		Transverse		Longitudinal	
			H	V	H	V	H	V	H	V
2.0	–30	–32	–43	–33	–34	–38	–37	–32	–30	–32
6.0	–37	–40	–39	–43	–48	–50	–35	–37	–47	–48
16.0	–46	–50	–48	–50	–50	–51	–46	–48	–45	–50

within an antenna test chamber. These procedures are outlined in detail in Chapter 9. The usual procedure is to set the probing system in the middle of the test region, set the probe antenna to a given angle, and then move the probe across the width of the test region. Knowing the maximum signal level with the probe looking at the source, the level drop when the antenna is clocked to its test angle, and the peak-to-peak ripple superimposed on the probe pattern, the extraneous signal level (reflectivity) can be found from Figure 6.6. The probe is generally clocked from 0 to 90 degrees in 10- to 15-degree steps. A transverse probe is recorded at each angle. This is done for both polarizations and for each agreed-upon test frequency. Next, the probe is set up to run down the axis of the chamber. The data are taken from 90 to 180 degrees, where the probe is looking toward the back wall at 180 degrees. Testing is normally done in an empty anechoic chamber. If an antenna test positioner is present in the chamber, the reflections from the wave diffracted from the positioner can contaminate the measurement, especially at the low frequencies where the antennas do not discriminate against local reflections.

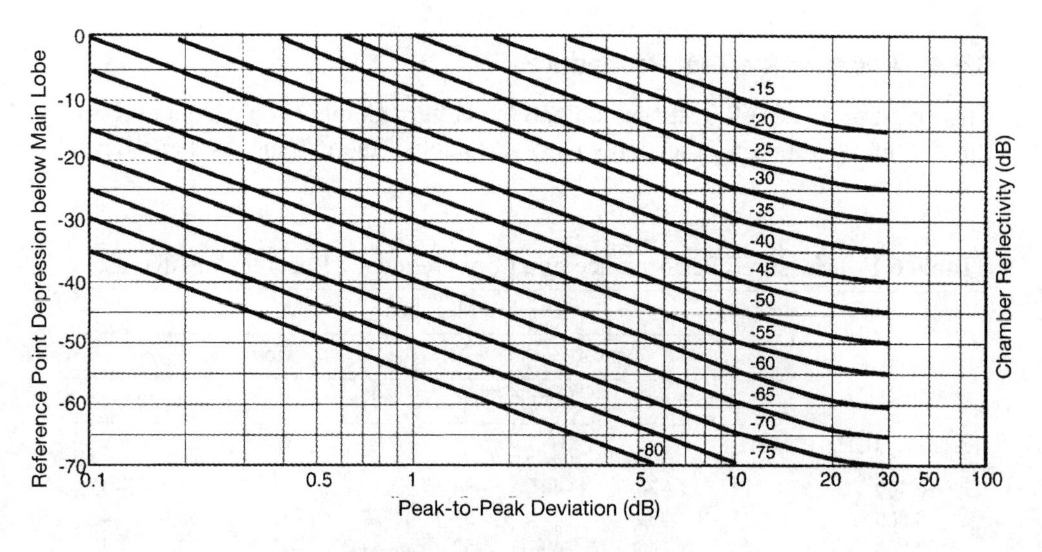

Figure 6.6 Reflectivity chart.

In actual use of the chamber, absorbers are used to cover the test positioner to minimize diffracted energy.

If the level of extraneous signal level in the chamber is critical, then a more thorough test is required. Recent advances in the field of near-field measurements have also made it possible to map an anechoic chamber's sources of reflections using an X–Y scanner and image software [2]. The process is further described in Chapter 9.

6.3 RADAR CROSS-SECTION TESTING

6.3.1 Design Considerations

The object of radar cross-section (RCS) testing is to determine the level of scattering that occurs when a test object is illuminated by radar. Two types of radar are used in tracking objects, (1) monostatic and (2) bistatic. A single radar site is used in the former, such as search radar on an aircraft, ship, or other mobile platform. The purpose is to provide target detection at the longest range possible so that a defensive response can be implemented prior to the enemy reaching their target. It is the task of the low observable design engineer to reduce the detectability of a military vehicle. To prove out the design concepts, radar cross-section measurements are necessary. In a monostatic RCS chamber, the design goal is to reduce forward scattering from reaching the device under test and also to reduce the amount of chamber backscatter, which raises the background level of the test region. These chambers are generally rectangular in geometry. In bistatic radar testing, the chamber must be designed such that the chamber background level is within certain limits. Critical placement of the various types of absorbing material determines whether the design achieves the above goals. Far-field bistatic chambers are generally square in shape with the test region located in the center of the enclosure. The absorber design procedure is detailed in Chapter 5.

RCS chambers have an additional requirement to provide a low equivalent RCS, also known as the chamber background level. This is required to maintain a good signal-to-noise ratio during RCS measurements and is determined by using the following analysis developed by Robinson [3].

First, the RCS of a flat reflecting metal sheet of known size is determined. This result is then applied to a computer model of the chamber walls using the wall reflectivity to develop the equivalent RCS of the anechoic chamber. This equivalent RCS is then assumed to occur at the location of the test region providing the CW background level. In chambers that are instrumented with hardware gated (pulse) radar, the back-wall reflected energy is gated out, thus improving the signal-to-noise performance in the test region.

Energy leaves the illuminating antenna and is incident on the absorber material. A portion is absorbed and the remainder is reflected. The antenna receives a portion of the reflected energy. We assume that the reflection is specular (i.e., reflected as if from a flat surface) and is not diffracted by any periodicity of the sur-

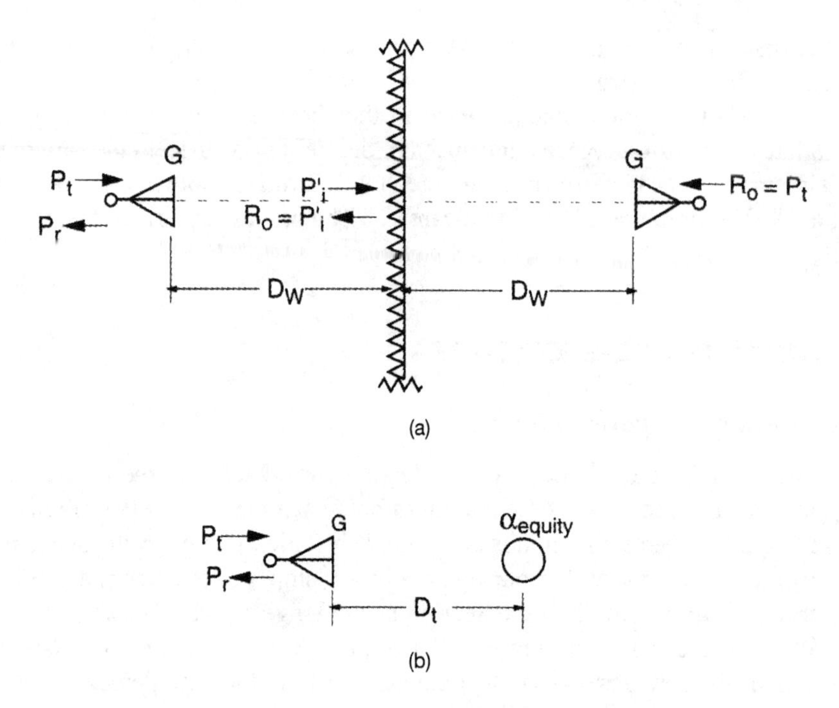

Figure 6.7 Method of images. (a) An antenna illuminating a wall. (b) Equivalent RCS.

face. It is well known [4] that the ratio of power received to that transmitted can be calculated by the method of images. The received signal is assumed to be transmitted from the image of the actual antenna (drawn in dashed lines in Figure 6.7). If the wall had a 100% reflection coefficient, the free-space propagation relation from the image to the antenna would be

$$(P_r/P_t) = G_1^2\lambda^2/((4\pi)^2)(2D_W)^2) = G_1^2\lambda^2/(64\pi^2 D_W^2) \tag{6-1}$$

where

$(P_r/P_t)_1$ = ratio of received to transmitted power
G_1 = antenna power-gain ratio (dimensionless) in the direction to the wall (not necessarily in the direction of the antenna axis, as implied in Figure 6.7)
λ = free-space wavelength
D_W = distance to the wall (i.e., the partially reflecting flat sheet) in the same units as λ

Because the wall only reflects part of the wave incident on it, we have for the actual situation

$$(P_r/P_t)_1 = R_o G_1^2\lambda^2/(64\pi^2 D_W^2) \tag{6-2}$$

where

R_o = the ratio at the wall of the reflected power density (W/m^2) to the incident power density for a normally incident wave. The usual parameter used to describe the performance of microwave absorber is $10\log_{10}(R_o)$, in decibels.

Another standard relationship is the radar equation [5], written as

$$(P_r/P_t)_2 = G_o^2\lambda^2\sigma/((4\pi)^3 D_t^4) \tag{6-3}$$

where

σ = the RCS of a radar target in square meters
D_t = the radar range to the target
G_o = peak antenna gain (assumed to be in the direction of the target)

The equivalent RCS of the wall is defined by determining the identical target size placed where the actual target will be located and give it the same received signal as the reflection from the wall. Equating the right-hand sides of equations (6-2) and (6-3) and solving for σ yields equations (6-4) and (6-5):

$$\sigma_{equiv} = \pi R_o(G_1/G_o)^2(D_t^4/D_W^2)m^2 \tag{6-4}$$

or

$$\sigma_{equiv} = 10\log\pi + 10\log R_o + 20\log(G_1/G_o) + 40\log D_t - 20\log D_w \text{ (dB}_{\text{sm}}) \tag{6-5}$$

Note that D_t and D_W should be in meters to give σ_{equiv} in the usual units of square meters or decibels relative to one square meter (dB_{sm}) (the logarithm is to the base 10). Note that wavelength and absolute antenna gain drop out of the final result. For the wall directly behind the target in the anechoic chamber, the ratio (G_1/G_o) usually is unity, because the antenna is directed at the target and is normal to the back wall. For other walls the ratio is usually significantly less than unity. The ratio appears squared because the pattern shape comes into play twice: during transmission and during reception. Also note that the numerical value assigned to σ_{equiv} is a strong function of the distance D_t at which the target will be placed.

The above analysis was based on an infinite flat sheet, whereas an actual chamber has finite walls. The results are still applicable to a rectangular chamber with 90-degree corners. The walls act as images to each other, and the result is that the walls appear as infinite as illustrated in Figure 6.8. Using the above analysis on a proposed chamber design, it was determined that only the wall behind the target contributed significantly to the residual RCS of the chamber. To minimize this contribution to the residual RCS of the chamber, several installations have used tilt walls or angled walls with good success. Because the tilted wall behaves as a very large aperture, the energy reflected is highly focused in the specular direction. Because the energy reflected from the tilted wall is not directed toward the receive antenna, multiple reflections occur and the energy is highly attenuated. Improvements on the order of 20–30 dB have been achieved.

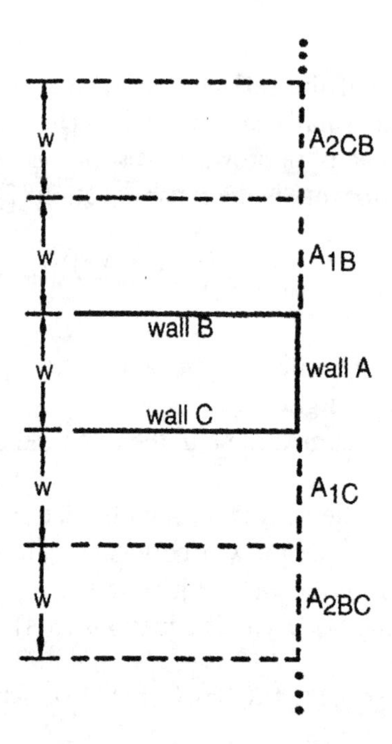

Figure 6.8 Imaging a wall into an infinite flat sheet.

Another design consideration of RCS chambers is the need for physical stability when the instrumentation radar uses background subtraction. The chamber wall stability needs to be on the order of 0.1 wavelength at the highest operating frequency. Again, the wall behind the target is the most critical in achieving good background subtraction performance.

6.3.2 Design Example

Assume that a far-field RCS range is required to test a target 0.46 m (1.5 ft) in diameter at 12 GHz. Further assume that the chamber is instrumented with a hardware gated pulse measurement radar. The minimum range length suitable for RCS testing is $2D^2/\lambda$ [6]; thus the range length should be on the order of 14.9 m (49 ft). At 2 GHz the test region would be on the order of 1.5 m (5 ft). To ensure that the angle of incidence for the side-wall absorber is on the order of 60 degrees, the chamber height and width should be on the order of 10.7 m (30 ft). The total overall length should be about 19.5 m (64 ft). This range length is required to allow for a 10-ns pulse signal gate around the target and another 10 ns to the back wall for the gate to have room to cut off in order to minimize the amount of reflected energy from the back wall falling within the gate [7]. It is assumed that the chamber will be used for measurements over the 2- to 18-GHz ranges, a common requirement. If the minimum reflectivity at 2 GHz is permitted to be –40 dB, then the back-wall absorber required is 0.46 m (1.5 ft) thick. If 0.79 m (2 ft) material is chosen, then the expected performance will be –44 dB minimum. The transmitter wall is chosen to be 0.30 m (1 ft) thick. Keep in mind that we will need to minimize any backscatter. It is necessary to select material that will guide the reflected energy into the back wall. Wedge material has been demonstrated to have lower backscatter than pyramidal material [8]. As a result, it has become quite common for anechoic chambers dedicated to RCS testing to have the bulk of the side walls, ceiling, and floor covered with wedge material. It is recommended that this material be of sufficient thickness to provide a minimum of –40 dB at 2 GHz. Also do not forget to allow for the directivity of the source antenna as discussed in Section 5.2. Drawing on the data from Figure 6.2, the absorber should be on the order of four wavelengths thick, or 0.61 m. The side-wall layout for this design example is shown in Figure 6.9. Note that it is recommended that patches of pyramidal absorber be located at right angles to the test region. These patches are required to terminate the energy scattered off the target in these directions, because they can fall within the gate of the instrumentation radar. Currently, the free-space chamber is not the range of choice for RCS measurements. The compact range has become the range design most commonly used for RCS measurements as discussed in Chapter 7. However, should a free-space RCS chamber be required, a detailed design approach is given in Robinson [3] for chamber reflectivity and chamber background RCS.

6.3.3 Acceptance Test Procedures

The acceptance testing of free-space RCS chambers is conducted using two different procedures. The procedure that most simulates the use of this chamber is to

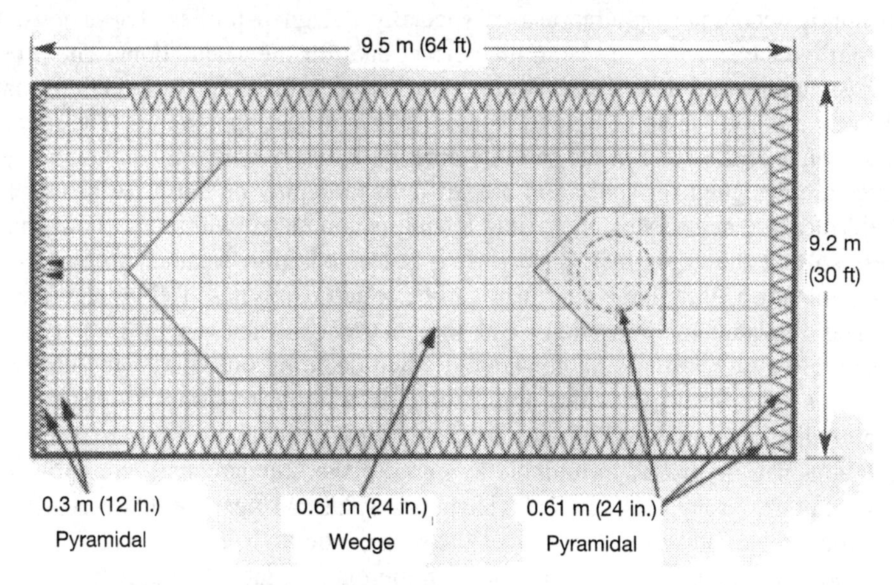

Figure 6.9 Side-wall absorber layout in a rectangular radar cross-section chamber.

take a known target, such as a sphere, and move it through the test region. This is accomplished by using either a translation mechanism or a rotator, with the sphere set up on a foam column offset from the center of rotation. As the sphere is moved through the test region, the radar return from the target will vary, depending on the ripple in the illuminating field. The peak-to-peak ripple can then be related to the level of extraneous signal level in the test region. The second procedure is to use a corner reflector target that is drawn through the test region with a translation device. The pattern of the corner reflector is quite broad, and the resulting signal variation is due to the extraneous signal phasing in and out with the direct path signal. The peak-to-peak variation can be used to determine the level of the extraneous signals.

6.4 NEAR-FIELD TESTING

6.4.1 Introduction

Because the recent availability of reasonably priced high-speed computers that have large amounts of memory, the near-field testing of antennas has become widespread. The process involves sampling the field near the face of the radiating aperture and then using a near-field to far-field transform to compute the far-field properties of the aperture. Several different geometries are utilized. The most common is the planar scanner that samples the field in an X–Y fashion and is used mostly for antennas that incorporate reflectors in their construction [9]. Another form rotates the antenna and samples the field in the vertical plane, which creates a cylinder of sampled data. A third form consists of sampling data in a spherical format, using various types of antenna positioning equipment. The sampling technique requires use of specially designed probes. These must be carefully characterized to obtain the correct data for the calculations. The procedures are software-driven and require very fast measurement equipment to make the test method viable compared to more conventional approaches. The method has seen extensive use in the satellite antenna industry where the antenna structures are designed for space and are difficult to deploy on earth for proper far-field measurements. Most near-field testing is conducted within anechoic chambers so that the equipment under test is protected from weather problems and contamination. Very large apertures can be tested using near-field techniques.

The fundamental limitations and the inherent advantages of each type of near-field system should be considered when deciding on what near-field technique is best for a particular application. It is also important to realize that seemingly "insignificant" limitations like having to deal with cable routings and rotations can be just as formidable as some of the "fundamental" limitations of a technique. From a theoretical viewpoint, spherical near-field measurements are the purest and most attractive of the three options. It is fairly probe-insensitive, low-cost, and easy to build from commercial available equipment, and it allows one to measure any type of antenna. However, for testing large gravity-sensitive antennas, the movement of the antenna under test becomes restrictive. Also, the data processing is significantly more complex than that for planar near-field testing.

Cylindrical near-field testing requires single axis rotation of the antenna under test. This may have significant advantages for testing in certain instances. This type of testing is ideally suited for base station personal communication system (PCS) antennas (antennas that radiate in an omnidirectional fashion in one plane with little energy radiated upwards or downwards).

Planar near-field testing is used for high directivity antennas (>15 dBi). The main attraction of this measurement technique is that the antenna under test remains stationary during the testing. For large spacecraft antennas, this is often the only feasible approach. Planar near-field testing is more intuitive than the other techniques. Data processing is simpler. The alignment procedures are easier to implement. In an antenna market where flat conformal antennas are becoming more popular, this technique is likely to be used extensively.

The physical arrangements for the various near-field measurement techniques are illustrated in Figures 6.10a–6.10c.

Table 6.3 compares the three standard near-field measurement techniques.

6.4.2 Chamber Design Considerations

The most important consideration in near-field anechoic chamber design is to ensure that the radiating aperture is properly terminated. That is, the principal beam of the antenna must see a load termination of sufficient attenuation that the energy reflected is sufficiently down with respect to the direct energy. The sampling method must not be degraded due to extraneous energy. In the main beam of the antenna, the field strength is high. It is relatively easy to minimize the effects of extraneous energy. However, as the probe scans off axis and the signal level drops, greater care is required to maintain a sufficient difference of signal to extraneous energy to ensure that the measured level is not degraded beyond acceptable levels. It is recommended that the extraneous signal level be at least 60 dB below the measurement level. This will ensure that the chamber has negligible effect on the measurements. Space loss from the emitter to the reflecting surface and back to the probe aids in reducing the effect of the chamber reflecting surfaces. Typically, the absorber should have a reflectivity on the order of 30–40 dB or better to achieve the overall required attenuation. The space loss to the nearest reflecting surface can be calculated using the following formula:

$$\text{Space loss (dB)} = 20 \log (4\pi R/\lambda) \tag{6-6}$$

where

R = distance from face of antenna under test to reflecting surface and back to the probe antenna on the near field scanner

λ = operating wavelength

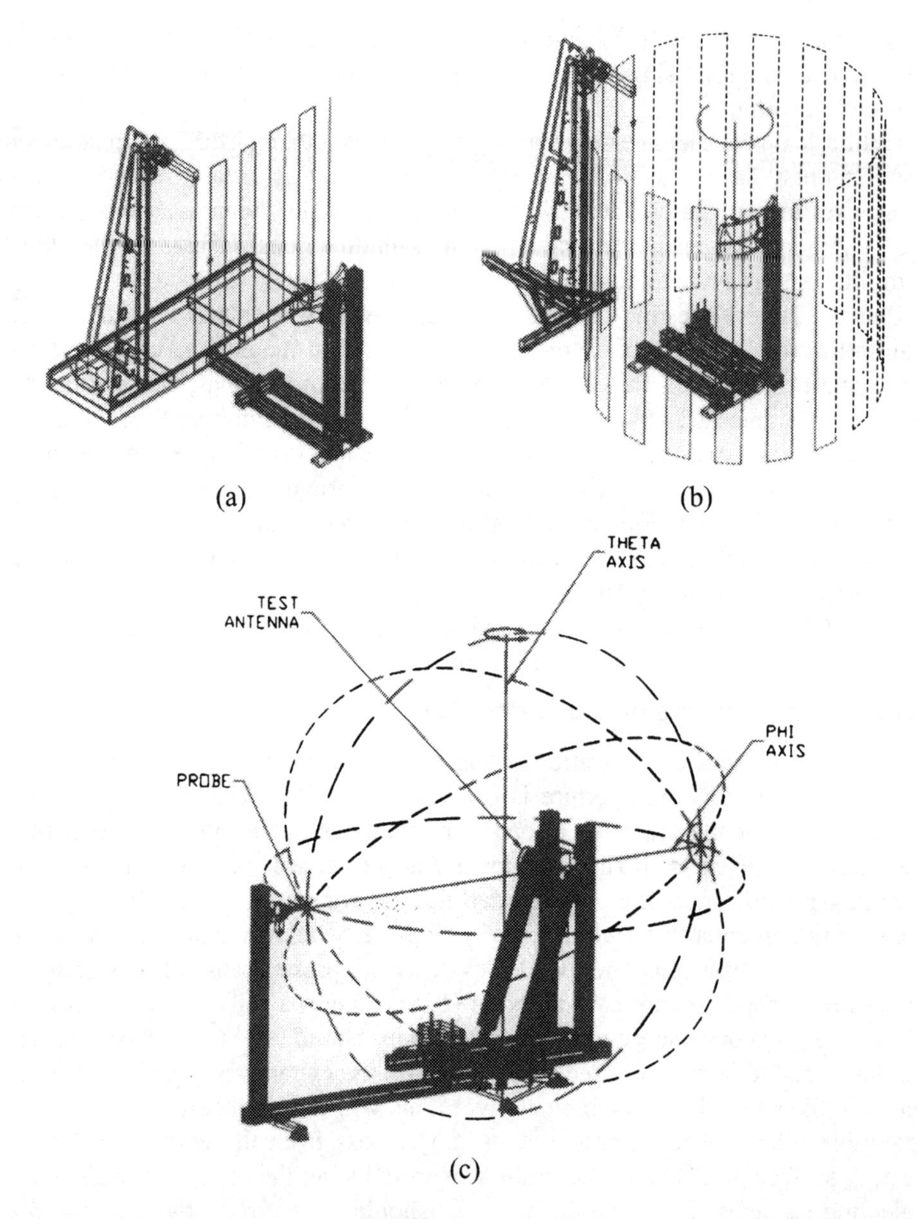

Figure 6.10 (a) Planar, (b) cylindrical, and (c) spherical near-field test arrangements.

6.4.3 Design Example

For a planar near-field test facility, Newell [10] recommends an error estimate of –0.001 dB. This translates into a combination space loss and absorber reflectivity of –60 dB. Knowing the distance from the antenna under test (AUT) and the reflecting surfaces adjacent to the near-field probe and the distance to the probe, the

Table 6.3 Comparison of Near-Field Measurement Techniques

Antenna Type/Parameter	Planar	Cylindrical	Spherical
High-gain antennas	Excellent	Good	Good
Low-gain antennas	Poor	Good	Excellent
Stationary AUT	Excellent	Possible	Possible
Zero-gravity simulation	Excellent	Poor	Variable
Alignment ease	Simple	Difficult	Difficult
Speed	Fast	Medium	Slow

space loss can be calculated and the remainder of the budget would be the reflectivity requirements of the absorber located on the reflecting surfaces.

6.4.4 Acceptance Test Procedures

The pattern comparison procedure described in Chapter 9 is recommended for evaluating near-field ranges. The procedure calls for multiple antenna patterns to be taken of the same antenna. After each data run, the antenna is moved with respect to the scanning mechanism so that the phase relationship is changed as least by $\lambda/8$. After the data are taken, the far-field patterns are calculated and plotted. The patterns are normalized to the peak of the pattern by overlaying the patterns. The differences in peak-to-peak variations at the various pattern levels are noted. The chart in Figure 6.6 is used to determine the extraneous signal level present in the test aperture. In cylindrical and spherical near-field ranges, the directive nature of the antenna can be used to locate sources of extraneous energy from the chamber surfaces. Improvements can then be implemented until the levels of extraneous signal are acceptable. A detailed procedure for evaluating near-field range multipath is given in Ref. 11.

6.5 ELECTROMAGNETIC COMPATIBILITY TESTING

6.5.1 Introduction

In recent years, most new anechoic chamber construction has been for radiated emission and immunity testing. The proliferation of electronic devices has led to problems of electromagnetic interference. As a result, governmental regulation for the amount of unintended emissions a device can emit over certain frequencies has evolved. In addition, some governmental agencies are requiring that electronic devices be immune to tolerable levels of radiated energy over a broad frequency spectrum. In the United States, the Federal Communications Commission (FCC) leads this effort by setting emission standards based upon measurements conducted on an open-area test site (OATS). Three measurement range lengths are at 3, 10, and 30 m. Because the frequency range of interest is 30–1000 MHz, measurements had to be conducted over a ground plane. Thus, it was necessary to allow for the ground reflection. The scan height of 1–4 m was chosen empirically. It was determined that for the 30- to 1000-MHz range and for the DUT and antenna separations, the "ripple" from the ground reflections are minimized using this scan range reducing the errors to acceptable values. A site attenuation requirement was determined so that the emission levels could be computed from the levels measured. Antenna calibration procedures [12] were established and emission limits were set. Outdoor measurements can be difficult to obtain, depending on (a) the location of the facility, (b) weather, (c) high ambient signals, and (d) location of the facility with respect to the production facilities. Soon, the larger companies were building indoor test facilities with the anechoic materials that were available at the time. The current technology uses a hybrid ferrite absorber to achieve low-frequency performance. A variety of emission and immunity test procedures are conducted in these chambers [13].

6.5.2 Design Considerations

The design of an anechoic chamber for EMI measurements involves the reduction of reflected energy from the enclosure walls and the damping of cavity modes generated when electromagnetic energy is inserted into a cavity formed by the six-sided enclosure. Both of these effects cause distortions in the energy transmitted from an emitter and received by a measuring device. A secondary consideration is the type of antenna used during the measurements. Directional antennas, such as log-periodic dipole arrays or horn antennas, help reduce the energy that reaches the end walls during chamber calibration. The usual method of determining the suitability of a chamber for use in measuring emissions from a device under test is to determine the site attenuation between the emitter and the measurement point. This is a derived value based upon antennas with known properties and calibrated signal sources and measurement receivers. The process is described in the ANSI Standard C63.4-1992 [14]. Longer measurement path lengths increase the space loss. Higher loss requires better performance absorbing material to minimize the uncertainty in the measurements. Three different path lengths are now common: 3, 5, and 10 m. A desktop computer can be conveniently measured in a 3-m chamber, whereas an automobile requires a 10-m facility. The type of absorbing material needed for a given chamber design is basically determined by the path length and the frequency range over which the measurement is to be performed. Experience has shown that for a 3-m chamber, absorber performance must be at least –18 dB at 30 MHz rising to greater that –24 dB around 100 MHz. It can then fall below –18 dB at 1000 MHz, if broadband biconical and LPA antennas are used to calibrate the site attenuation and to perform the measurements.

6.5.3 Design Examples

6.5.3.1 Three-Meter Test Facility. This range length is the most common for anechoic facilities built for emission testing. Typical size is 9.5 m long, 6.5 m wide, and 5.7 ms high, with the latter allowing antenna elevation to 4 m above the ground plane. When operating over the 30- to 1000-MHz frequency range, ferrite tile is used to cover the chamber surfaces and is mounted on ½-inch-thick plywood. If the measurements are extended to 18 GHz, then special dielectric foam absorbers are added to the face of the ferrite tile. Both flat- and grid-type ferrite tile are used in these installations. Full correlation in accordance with ANSI C63.4 and EN 50147-2 are possible in this size chamber. These chambers are available from a number of anechoic chamber manufacturers. With the addition of absorber on the floor of the chamber, full compliance with IEC 61000-4-3, for immunity testing, can be achieved. Examples of these types of chambers are discussed in Chapter 10.

6.5.3.2 Ten-Meter Test Facility. For larger equipment, the 10-m test facility is often required. A full FCC registered facility at 10 m requires that the chamber be on the order of 19 m long, 13 m wide, and 8.5 m high. These chambers use hybrid absorbers that have been optimized for operation over the full 26-MHz to 18-GHz frequency range. This size chamber is suitable for ANSI C63.4, CISPR 16, EN 55022, EN 50147-2, and similar international regulations. With the addition of a 3-m floor patch, the chambers can also be used for EN 61000-4-3 immunity testing.

6.5.3.3 The Fully Anechoic EMC Chamber. There is a movement within the European community to use a 3-m fully anechoic chamber as the best overall solution for radiated emission testing [15, 16]. These chambers eliminate the requirement for scanning from 1 to 4 m, as required in five-sided chambers with a ground plane. The result is easier calibration and faster measurements. The design is the same for a five-sided 3-m chamber, with the addition of absorber on the floor. The entire floor should be covered, as is in a microwave chamber. Provisions for walkways to the test stand should be provided so that the item under test can be easily moved in and out of the chamber. There is also a move within the industry to develop multifunctional test chambers, as described in Ref. 17.

6.5.3.4 MIL-STD-461 Requirements. MIL-STD-461 [18] is the general specification for emission and susceptibility requirements for military equipment. It requires that the testing be conducted within a shielded enclosure. RF absorber material (carbon-impregnated foam pyramids, ferrite tiles, and so forth) should be used when performing electric field radiated emissions or radiated susceptibility testing inside a shielded enclosure to reduce reflections of electromagnetic energy and to improve accuracy and repeatability. The RF absorber is to be placed above, behind, and on both sides of the equipment under test and behind the radiating or receiving antenna. The minimum normal incidence performance of the absorber is to be 6 dB from 80 MHz to 250 MHz, as well as 10 dB above 250 MHz.

6.5.4 Acceptance Test Procedures

EMC chambers involve measurements that prove the device under test meets or exceeds a specific emission or immunity level. Therefore, the calibration of such facilities must meet specified calibration procedures. In the United States, the Federal Communications Commission sets the requirements. The procedures in the FCC documents must be used to qualify a particular test site. Alternative sites, such as anechoic chambers, require that the site be evaluated using a procedure known as volumetric site attenuation.

The acceptance test procedures for emission chambers are called out in the various EMC specifications. Generally, the procedures are similar to those specified in ANSI C63.4-1992. This specification requires semianechoic chambers to have normalized site attenuation measurements performed using volumetric test procedures. This involves testing at five locations within the test region and at two heights. All of the site attenuation curves must fall within ±4 dB of theoretical open-area test site (OATS) values. This measurement is very critical to the overall accuracy of the emission measurements. The most critical component of the calibration procedure is the accuracy of the antenna calibrations. It is recommended that the antennas be calibrated on an OATS using the three-antenna method [19].

6.6 IMMUNITY TESTING

6.6.1 Introduction

Immunity chambers compliant with IEC 61000-4-3 or its following documents require that the field be uniform within 0 to +6 dB over a 1.5 × 1.5-m aperture located 0.8 m off the chamber floor, at a range length of 3 m. This is tested by locating a field sensor at a point within the aperture and normalizing the field at that point to the level that is to be tested, such as 3 or 10 V/m, over the 80-MHz to 1000-MHz frequency range (EN 61000-4-3). The remaining 15 points in the aperture are tested and plotted. The resultant plot must demonstrate that 75% of the points fall within the 0- to + 6-dB requirement.

Immunity testing requires the use of a shielded enclosure to prevent interference with adjacent electronic systems during testing of the device. If shielded enclosures are used in the construction of a 3-, 5-, or 10-m emission chamber, then simple addition of an absorber patch on the floor permits the chamber to perform to the immunity requirements of IEC 61000-4-3, or similar requirements. The mode-stirred chamber is another type of chamber that has been developed, which is nonanechoic, but has been found to be an efficient means of conducting immunity testing.

6.6.2 Mode-Stirred Test Facility

A nonanechoic approach to radiated emission and immunity testing has been developed which, for the sake of completeness, is now presented [20]. The concept is not to reduce the effect of cavity standing waves using absorption material, but to reinforce the standing waves by designing the chamber to have as high a Q as possible, by minimizing wall losses. A mode-stirring device (paddles) is inserted in the chamber to change the mode structure at a given point in the chamber. This allows for an average field strength to be generated over the chamber's operating bandwidth, illustrated in Figure 6.11. These test facilities are commonly called "statistical mode-averaging reverberation test sites" [21]. Unlike anechoic chambers, which use absorbers to minimize chamber modes, mode-stirred facilities use the internally reflected energy from the surfaces to create resonant cavity modes. These modes are "averaged" or "stirred" within the chamber by changing the boundary conditions of the cavity. The boundary conditions are changed in the chamber by moving strategically placed, reflective paddles, which have surfaces that are large with respect to the chamber size and wavelength of test operations.

Mode-stirred chambers are constructed from a variety of materials. The most common is welded steel or aluminum, although modular style chambers having galvanized steel surfaces are also used. The inside finish of these facilities can be

Figure 6.11 Mode-stirred reverberation chamber for measuring electromagnetic immunity of automobiles. The chamber is constructed of prefabricated shielding. Note the mode-stirring device in the upper right corner of the photograph. The LPA shown to the left of the vehicle is used to excite the chamber modes. [Photograph courtesy of EMC Test Systems (ETS), Austin, TX.]

custom-treated for different Q values. The Q determines the property of the chamber response in terms of mode formation. One of the big advantages of the mode-stirred chamber is the power density possible for a given input power. A 1-Watt input can provide average power densities on the order of 0.25 mW/cm^2. The calibration procedures for these chambers is extensive, and the manufacturer must have the necessary skills and equipment.

6.7 EM SYSTEM COMPATIBILITY TESTING

6.7.1 Design Considerations

Chambers used for system compatibility are generally large shielded enclosures lined with a uniform treatment of absorber, described in Chapter 10. The primary consideration is suppression of reflected energy off the chamber surfaces, in order to achieve adequate isolation between various antennas located on the device under test, such as multiple antenna systems on an aircraft. The reflection loss required of the absorber is determined by first calculating the path loss between the antennas under test and the closest chamber wall. This is given by

$$PL(\text{dB}) = 20\log(4\pi R/\lambda) \tag{6-7}$$

The amount of isolation required is generally on the order of 100 dB. The difference between the path loss and 100 dB determines the reflectivity required of the absorber at the frequency of interest. The size of the absorber required to achieve the required reflectivity is determined by referring to the absorber specifications in Chapter 3.

6.7.2 Acceptance Testing

Assume that it is required to test isolation between antennas on a spacecraft. The system operates in the 2- to 4-GHz frequency range. The distance to the chamber side walls is 12.2 m. Isolation required between the antennas is 100 dB. In order to resolve 100 dB, the dynamic range of the measurement system must exceed 106 dB. Assume that the receiving system (spectrum analyzer) has –90 dBm sensitivity at 2 GHz. Also assume that the two antennas have 16-dBi gains, then the transmitter should have a minimum output of –16 dBm to achieve the required dynamic range. The antennas are focused so that the angle of incidence is equal to the angle of reflection at the absorber-covered wall. A reference level is taken by placing a large flat metal sheet at the surface of the absorber, and a reference level is set in the test instrumentation. The reflector is removed, and the measured reflectivity is the difference. This procedure is repeated over the entire wall area of the chamber at a sufficient number of locations to get a good representation of the performance of the wall absorber in the chamber.

REFERENCES

1. J. Appel-Hansen, Reflectivity Level of Radio Anechoic Chambers, *Transactions of Antennas and Propagation, Vol. AP-21,* No. 4, July 1973.
2. G. Hindman, and D. Slater, Anechoic Chamber Diagnostic Imaging, *Antenna Measurement Techniques Association Proceedings* 1992.
3. L. A. Robinson, *Design of Anechoic Chambers for Antenna and Radar-Cross-Section Measurements,* AD-B074391L, November 1982.
4. S. Silver, *Microwave Antenna Theory and Design,* MIT Radiation Laboratory Series, Vol. 12, Section 15.20, McGraw-Hill, New York, 1949.
5. E. F. Knott et al., *Radar Cross Section,* 2nd edition, Artech House, Boston, 1993.
6. R. G. Kouyoumjian, and L. Petera, Jr., Range Requirements in Radar Cross-Section Measurements, *Proceedings of the IEEE,* Vol. 53, No. 8, pp. 920–928, August 1965.
7. D. W. Hess and V. Farr, Time Gating of Antenna Measurements, *Antenna Measurement Techniques Association Proceedings,* pp. 1-9–1-17, 1988.
8. S. Brumley, *Anechoic Chamber Simulation and Verification,* California State University Northridge, June 1993, ADA308236.
9. D. Slater, Near-Field Test Facility Design, *Antenna Measurement Techniques Association Proceedings,* pp. 3-1–3-10, 1985.
10. A. C. Newell, Error Analysis Techniques for Planar Near-Field Measurements, *IEEE Transactions on Antennas and Propagation,* Vol. 36, No. 6, pp. 254–268, June 1988.
11. G. F. Masters, Evaluating Near-Field Range Multi-Path, *Antenna Measurement Techniques Association Proceedings,* pp. 7–15, 1992.
12. ANSI Standard C63.5-1992, *American National Standard for Methods of Measurement of Radio-Noise Emissions from Low Voltage Electrical and Electronic Equipment in the Range of 9 kHz to 40 kHz,* ANSI, 1992.
13. **EMC Standards by Source:**

    CENELEC Harmonized Standards

    EN 55011:1998—Industrial, scientific, and medical (ISM) radio-frequency equipment—Radio disturbance characteristics—Limits and method of measurements.

    European Standards

    CISPR11:1999—Industrial, scientific, and medical (ISM) radio-frequency equipment electromagnetic disturbance characteristics—Limits and methods of measurement.

    CISPR16-1:1999—Specifications for radio disturbance and immunity measuring apparatus and methods—Part 1: Radio disturbance and immunity measuring apparatus.

    CISPR22:1997—Information technology equipment—Radio disturbance characteristics—Limits and methods of measurement.

    EN 55011:1998—Industrial scientific, and medical (ISM) radio-frequency equipment—Radio disturbance characteristics—Limits and methods of measurement.

    EN 55022:1994—Limits and methods of measurement of radio disturbance characteristics of information technology equipment.

EN 61000-4-3:1996—Electromagnetic compatibility (EMC)-Part 4-3: Testing and measurement techniques—Radiated, radio-frequency, electromagnetic field immunity test.

United States Standards

American National Standards Institute (ANSI) C63.4:2000—Standard for methods of measurement of radio noise emissions from low-voltage electrical and electronics equipment in the range of 9 kHz to 40 GHz.

ANSI C63.5:1998—American national standard for electromagnetic compatibility—Radiated emission measurements in electromagnetic interference (EMI) control—Calibration of antennas.

United States Military Standards

MIL-STD-461E:1999: Requirements for the control of electromagnetic interference characteristics of subsystems and equipment.

Society of Automotive Engineers (SAE)

Standard: J551/1: Performance levels and methods of measurement of electromagnetic compatibility of vehicles and devices (60 Hz to 18 GHz).

Standard: J-1113/21: Electromagnetic compatibility measurement procedures for vehicle components, Part 21: Immunity to electromagnetic fields; 10kHz to 18 GHz, absorber lined chamber.

14. ANSI Standard C63.4-1992, *American National Standard for Methods of Measurement of Radio-Noise Emissions from Low Voltage Electrical and Electronic Equipment in the Range of 9 kHz to 40 kHz,* ANSI, 1992.
15. C. Vitek, Free-Space Chambers for Radiated Emission Measurements, *Evaluation Engineering,* pp. 150–161, May 1997.
16. R. A. McConnell, and C. Vitek, Calibration of Fully Anechoic Rooms and Correlation with OATS Measurements, *IEEE 1996 International Symposium on Electromagnetic Compatibility,* August 1996, pp. 134–139, Santa Clara, CA.
17. B. Kwon et al., A Multi-Functional Anechoic Chamber for Far/Near Field Antenna Measurements and EMC/EMI, *Antenna Measurement Techniques Association Proceedings,* Montreal, Canada, pp. 422–426, October 1998.
18. MIL-STD-461E:1999: *Requirements for the control of electromagnetic interference characteristics of subsystems and equipment.*
19. IEEE, *IEEE Standard Test Procedures for Antennas,* IEEE STD 149-1979, IEEE, New York, 1979.
20. M. Crawford and G. Koepke, *Design, Evaluation, and Use of Reverberation Chamber for Performing Electromagnetic Susceptibility/Vulnerability Measurements,* National Bureau of Standards, Boulder, Colorado, Technical Note 1092 (1986).
21. R. A. Zacharias and C. A. Avalle, Applying Statistical Electromagnetic Theory to Mode Stirred Chamber Measurements, *Engineering Research, Development, and Technology,* Lawrence Livermore National Laboratory, Livermore, CA, UCRL-X 53868-92, 723 (1993).

CHAPTER 7

THE COMPACT RANGE CHAMBER

7.1 INTRODUCTION

Measurements on microwave antennas are based on accurate electromagnetic environment simulation, so that any errors between the measured and true radiation pattern of a test item are minimal. The importance of high-quality measuring ranges is evident. Slight errors between measured and true radiation patterns could raise doubts as to the design technique or the theoretical approach used in predicting the radiation properties of the test item. This applies to both antenna and radar cross-section measurements.

Most antenna measurements need to be carried out in the far field; that is, the test antenna should be illuminated by a plane wave. The most frequently used method of achieving this is to provide sufficient separation between the source and the item under test so that the spherical wave approaches a plane-wave character.

A short range test method, invented by Johnson [1], is to place the equipment under test in the radiating near zone of a large reflector. Johnson determined that if the reflector was sufficiently large and the edges of the reflector were properly terminated, the field in the radiating near field (this region falls between the near field and far field of the antenna aperture) would be adequately uniform in amplitude and phase for electromagnetic testing. The energy off the reflector is collimated into a very focused beam with a uniform phase distribution on the axis of the reflector. This approximation of a far field can be accomplished by several methods. The most common compact range is the prime focus version, which uses an offset feed to illuminate the main reflector. The field uniformity is controlled by the design of the feed. The phase is controlled by the surface of the parabolic reflector. Another method is to use two perpendicular parabolic cylinders. A third method is to use a Cassegrain or Gregorian antenna where the main reflector is shaped to control the phase, and amplitude is controlled by a special feed illuminating a shaped subreflector. Merits of each of these systems and the design of the anechoic chambers housing the systems are now discussed.

Another means of forming a plane wave in a short distance is to use a microwave lens to achieve the uniform phase front. Chambers for housing these type compact ranges is essentially the same as the prime focus reflector system discussed herein.

These compact ranges make it possible to measure electrically large apertures within a moderately sized anechoic chamber.

7.2 ANTENNA TESTING

7.2.1 Prime Focus Compact Range

The range operation begins with a spherical wave being radiated by the reflector feed located offset from the reflector. The feed design provides for uniform illumination over the face of the reflector and the parabolic shape straightens out the phase so that a uniform field is reflected off the reflector. The test region is located two focal lengths away from the reflector and is about one focal length deep. The test region is typically half the width and height of the reflector. This is illustrated in Figure 7.1.

A typical list of features for these ranges, as generally specified, is listed in Table 7.1.

The chamber and absorber layout for the compact range specified in Table 7.1 for antenna testing must be such that it does not degrade the compact range reflector's performance. The recommended chamber size is 5.5 m (18 ft) high, 11 m (36 ft) wide and 12.2 m (40 ft) long. The chamber back wall must properly terminate the range. To have an insignificant impact on the test region, the absorber reflectivity should be at least 10 dB better than the range specifications at 2 GHz or –35 dB. Thus, the minimum terminating wall pyramidal absorber should be at least 0.30 m (1 ft) thick. If the range is to be used at 1 GHz, then the absorber should be at least 0.60 m (2 ft) thick. It is quite common to use this size compact range down to 1 GHz because of the large number of global positioning system (GPS) antennas now being developed. The next item is to select the side-wall, ceiling, and floor absorbers. Because the energy off the reflector is highly collimated, the energy level reaching the side wall is typically 20–30 dB down from the main beam. The pyramidal absorber on the walls, floor, and ceiling is generally sized to be half the thickness of the back-wall absorber. This treatment is carried back behind the reflector, although it is common to leave the area on the wall immediately behind the reflector bare of absorber, because very little energy

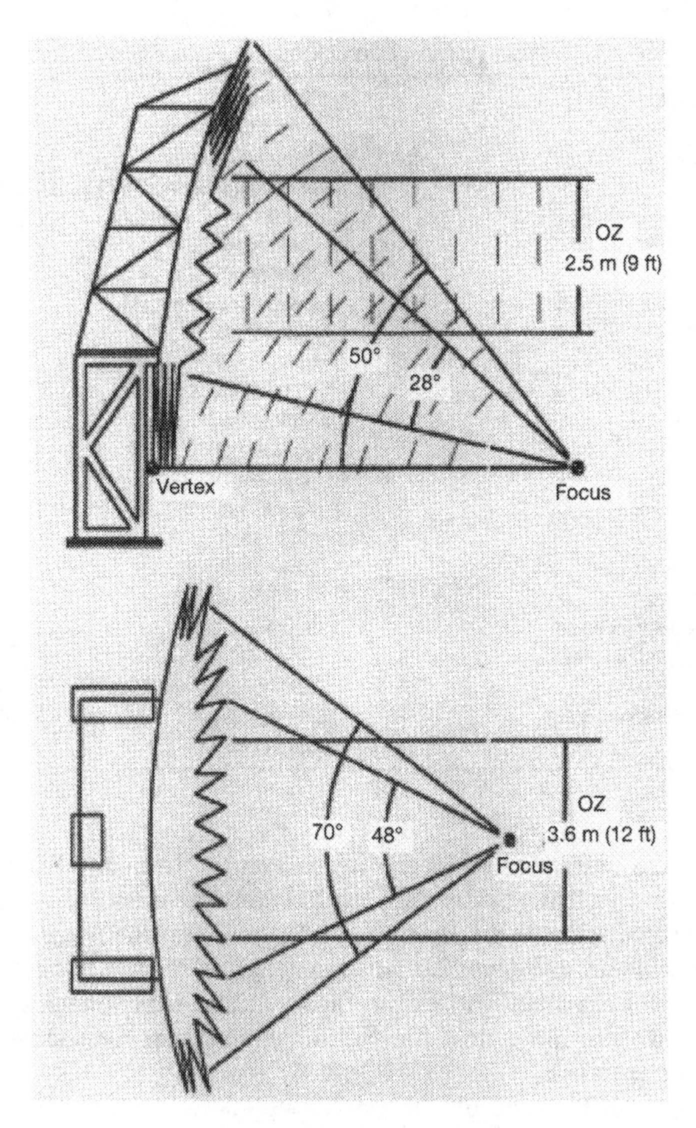

Figure 7.1 Operation of the prime focus compact range.

reaches this zone of the chamber. A variety of antennas are used to feed the prime focus compact range.

7.2.2 Dual Reflector Compact Range

The dual reflector compact range [2] is illustrated in Figure 7.2 using a scale model aircraft as the test vehicle. The system antenna feed is located on the floor and is focused on the parabolic subreflector located on the ceiling. This, in turn, is focused on the main reflector at the end of the room. This reflector collimates the energy down the center of the chamber to the test region. The reflector and feed

Table 7.1 Typical Prime Focus Compact Range Specifications

Frequency range:	2–94 GHz
Reflector size:	6.1 m (20 ft) wide, 4.3 m (14 ft) high
Test region size:	Elliptical cylinder
	H: 1.2 m (4 ft), W: 1.8 m (6 ft), L: 1.8 m (6 ft)
Amplitude taper:	1.0 dB over 1.8 m (6 ft)
Total Phase variation:	< 10 degrees, 2–18 GHz
	< 20 degrees, 18–94 GHz
Maximum extraneous signal level:	Specification/typical
1.7–2.6 GHz	–20/–25 dB
2.6–3.95 GHz	–20/–25 dB
3.95–5.85 GHz	–25/–30 dB
5.85–8.2 GHz	–30/–35 dB
8.2 –12.4 GHz	–35/–40 dB
12.4–18.0 GHz	–35/–40 dB
18.0–26.5 GHz	–30/–35 dB
26.5–40.0 GHz	–28/–35 dB
40.0–60.0 GHz	–28/–35 dB
60.0–94.0 GHz	–25/–30 dB
Typical cross-polarization:	–30 dB
Typical range mutual coupling:	< –45 dB
Reflector type:	Parabolic section with offset feed
Height of test region center:	2.4 m (8 ft)
Focal length:	3.6 m (12 ft)

system can also be rotated 90 degrees in the chamber. Normally, the antenna feed is located in a wall port in the adjacent control room, and the reflectors are mounted on the walls of the chamber. The operation of the range illustrated in Figure 7.2 is shown in Figure 7.3. This range has the first reflector mounted on the ceiling of the chamber. The feed antenna is located in the floor of the chamber.

Table 7.2 lists the properties of a typical dual reflector compact range as they are generally specified.

The purpose of the anechoic chamber design is to house the compact range without degrading its performance. For the dual reflector compact range, the critical back-wall design parameter is the 0.4-dB peak-to-peak amplitude ripple, which requires that the back-wall absorber be 0.60 m (2 ft) thick. This same thickness of absorber must be placed about the subreflector for about half the length of the room to terminate the feed energy. Because of the collimating performance of the main reflector, the remainder of the room is covered in 0.30 cm (1 ft) pyramidal absorber. Small areas immediately behind each reflector need not be covered, because the reflectors shadow these areas. Special treatment, with absorber around the feed antenna, is required to achieve adequate isolation to the test region. The remainder of the absorber can be 0.30 m (1 ft) pyramidal material. The feed antennas need to be selected to provide a minimum level of energy directed in the direction of the test region while maintaining proper illumination of the first reflector.

Figure 7.2 Dual reflector compact range. (Photograph courtesy of the Raytheon Company, El Segundo, CA.)

7.2.3 Shaped Reflector Compact Range

The shaped reflector compact range focuses the energy so precisely that practically all the beam is in the test zone and the test chamber has very little effect on the system performance. The design uses proprietary software to generate complex surfaces on two reflectors. The surfaces, coupled with a high gain feed system, act to very efficiently map the desired plane wave characteristics to the test zone. The precise control of spillover effectively decouples the chamber.

In the shaped reflector technique, high-gain ($\approx$24 dB) corrugated horns with 10-degree beamwidths transmit patterns that remain nearly constant across the frequency band, by phase compensation and aperture sizing. The shaped surface of the subreflector maps the field distribution to a uniform level across most of the reflector surface and then tapers the amplitude as it approaches the edge. Multiple special horn antennas are required to feed the shaped reflector

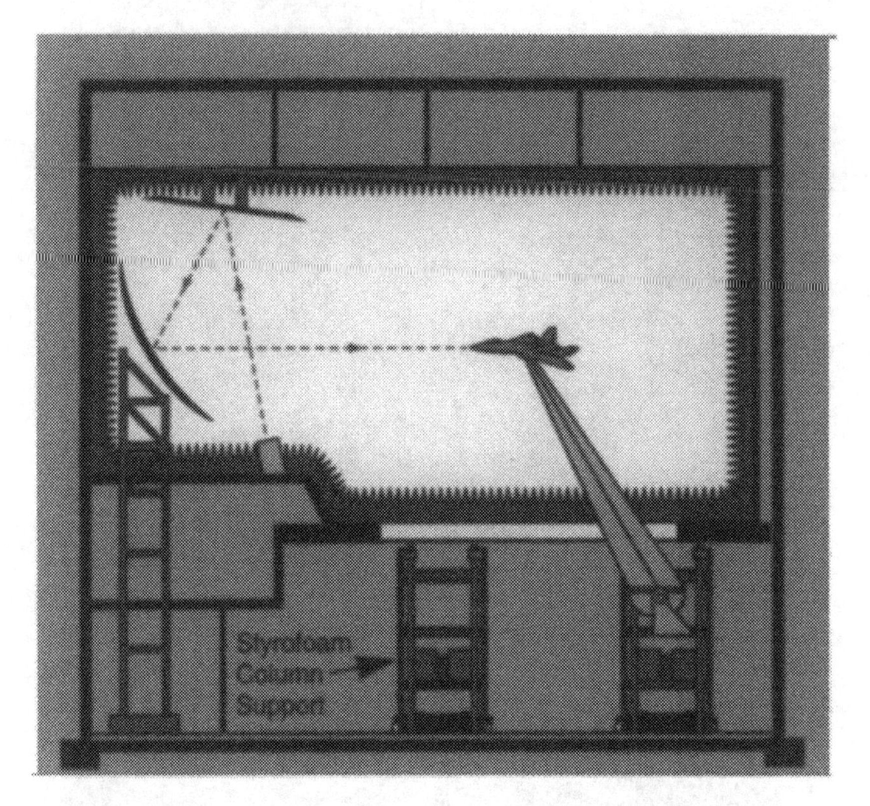

Figure 7.3 Operation of the dual reflector compact range.

compact range, as a result of the stringent beamwidth and phase requirements of the feeds.

The field distribution from the feed and subreflector illuminates the main reflector. The reflector surfaces are shaped to transform the fields to a plane wave in the test zone as in a prime focus system. The combined effect of both amplitude and phase shaping provides excellent plane wave performance in the quiet zone. The energy is highly focused in the shaped reflector compact range, as illustrated in Figure 7.4.

Results have shown that up to 75 percent of the linear dimension across the main reflector can be used as a test area. The driving consideration is ripple in the test zone, because a taper that is too sharp can behave as a scatterer. Using 50–60 percent of the linear dimension provides nominal ripple performance of less than 0.2 dB. As an example, one version of the system has a 12.2-m (40-ft) test region with an 18-m (59-ft) by 22.6-m (74-ft) elliptical main reflector.

Table 7.3 lists the typical characteristics of the Shaped Reflector Compact Range system.

The shaped reflector compact range is very highly collimated. Therefore, the primary consideration is proper termination of the main beam energy. To maintain the typical performance specified in Table 7.3, the 0.4-dB peak-to-peak amplitude

Table 7.2 Typical Dual Reflector Compact Range Specifications

Test zone dimensions	
Diameter:	1.6–2.4 m
Depth:	1.6–4.0 m
Frequency range:	2–100 GHz
Amplitude characteristics 1.6 m	
Total deviation:	1.30 dB (2–3 GHz)
1.0 dB (3–4 GHz)	
Taper:	0.40 dB (4–6 GHz)
	0.25 dB (6–40 GHz)
	0.40 dB (40–100 GHz)
Ripple (peak to peak):	0.60 dB (4–6 GHz)
	0.40 dB (6–40 GHz)
	0.60 dB (40–100 GHz)
Phase Characteristics 1.6 m	
Total deviation:	14 degrees (2–3 GHz)
	12 degrees (3–4 GHz)
Ripple (peak to peak):	6 degrees (4–6 GHz)
	4 degrees (6–40 GHz)
	8 degrees (40–100 GHz)
Cross-polarization (vertical plane):	0.8 m < –33 dB, 1.6 m < –27 dB
Mutual coupling:	–50 dB typical
Reflector dimensions:	Subreflector: 3.25 m, 3.8 m (W, H)
	Main reflector: 4.80 m, 3.8 m (W, H)
Recommended chamber size:	7.5 m wide, 4.9 m high, 13.2 m long

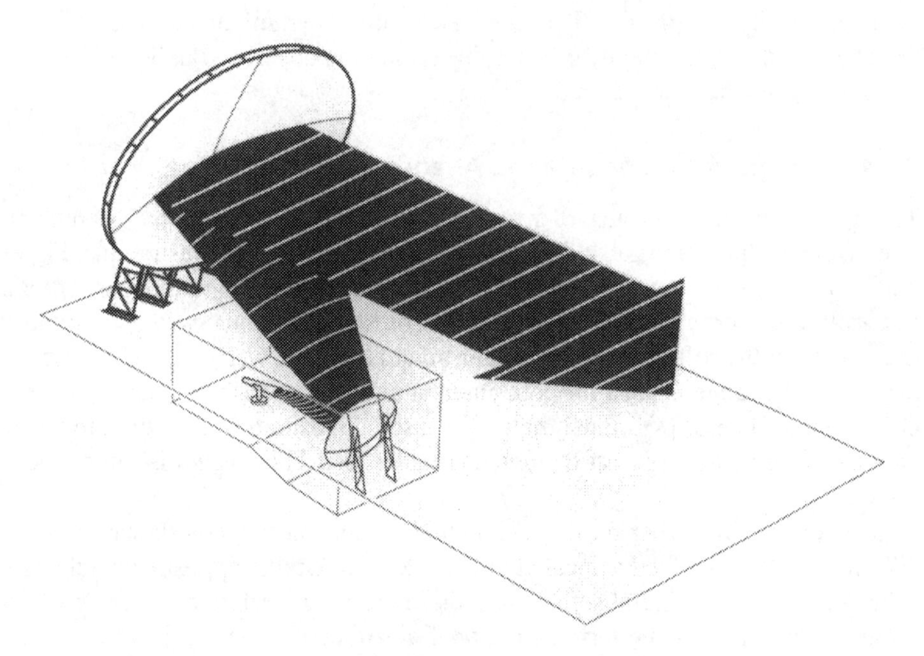

Figure 7.4 Operation of the shaped reflector compact range.

Table 7.3 Typical Shaped Reflector Compact Range Specifications

Test zone:	1.8-m (6-ft)-diameter cylindrical, 2.4 m (8 ft) long 6–95 GHz	
	1.2-m (4-ft)-diameter cylindrical, 1.8 m (6 ft) long 2–6 GHz	
Frequency range:	2–95 GHz	12.0–18.0 GHz
Feed antenna bands:	2–3.4 GHz	18.0–95 GHz, by request
	3.4–5.8 GHz	
	5.8–8.0 GHz	
	8.0–12.0 GHz	
Antenna VSWR:	2.0:1 max	
Extraneous energy:	–80 dB typical	
Performance:		

Frequency	Amplitude Ripple Peak (dB)	Amplitude Ripple Typical (dB)	Amplitude Taper Maximum (dB)	Phase Deviation Peak (degrees)	Phase Deviation Typical Deg.	Test Region Size, (meters)
2–4 GHz	1.0	0.4	2.0	10	6	1.2 × 1.2
4–6 GHz	1.0	0.4	1.0	10	6	1.2 × 1.2
6–8 GHz	0.8	0.3	0.8	10	4	1.8 × 1.8
8–40 GHz	0.7	0.2	0.8	10	4	1.8 × 1.8

Reflector size:	Main reflector: 4.2 m
	Subreflector: 1.93 m
Recommended chamber size:	6.1 m high, 7.6 m wide, 12.2 m long

ripple requires –33-dB signal attenuation at 2 GHz, plus 10 dB of margin, for a total reflectivity of –43 dB. This translates into a pyramidal absorber of three wavelengths thick, or roughly 0.46 m. The remaining absorber can be on the order of one-third as thick, or 0.16 m.

7.2.4 Compact Antenna Range Absorber Layout

The specific sizing of the absorber for the various compact ranges has been given. The layout of the various absorbers is shown in the following illustrations. Figure 7.5 illustrates the layout for the prime focus compact range. Pyramidal absorber is generally used throughout the chamber. In some installations, wedge material is used between the reflector feed mechanism and the back wall of the chamber, especially if the range is used for both antenna and RCS measurements. In antenna chambers, a patch of pyramidal material is used opposite the test region to terminate any reflected energy off the antenna under test. This design is illustrated in Figure 7.6.

The layout of absorber for the dual cylinder compact range is shown in Figure 7.7. In the figure the feed is located in the side wall. On the opposite wall the first reflector is located. The absorber on either side of the reflector must be of the same performance as the terminating wall absorber. The same absorber is used around the main reflector. The termination wall absorber is calculated as discussed in Section 7.2.2.

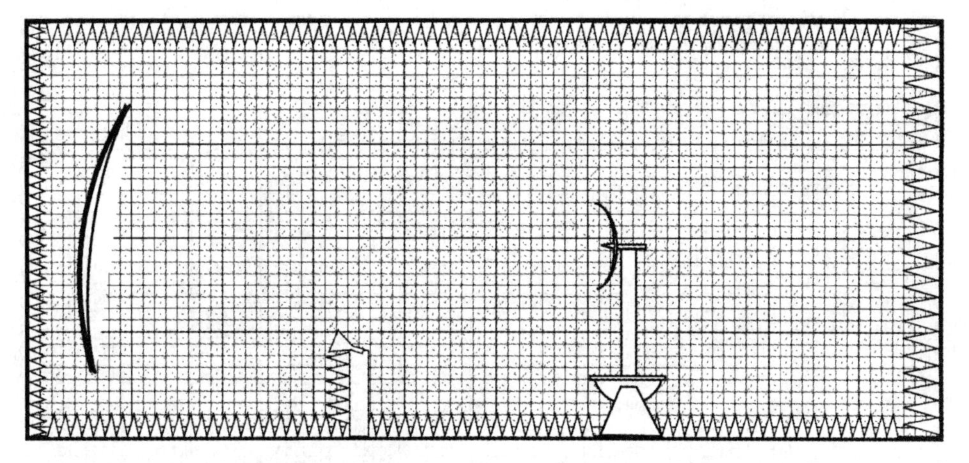

Figure 7.5 Absorber layout for prime focus compact antenna range.

7.2.5 Acceptance Testing of the Compact Antenna Anechoic Chamber

It is desirable to separate the acceptance test of the reflector system from that of the anechoic chamber housing the compact range. Generally, the manufacturer of the compact range reflector conducts a series of field probes to demonstrate the properties of the reflector after installation in the anechoic chamber. The chamber manufacturer has to demonstrate that his chamber provides a proper environment to house the compact range system. The best method to demonstrate this is to use the pattern comparison method described in Chapter 9. It is desirable to translate the positioner along the axis of the chamber to achieve the multiple pattern measurements required to perform the test method. It is recommended that a high gain antenna be used at each test frequency so that the sidelobes of the antenna are well

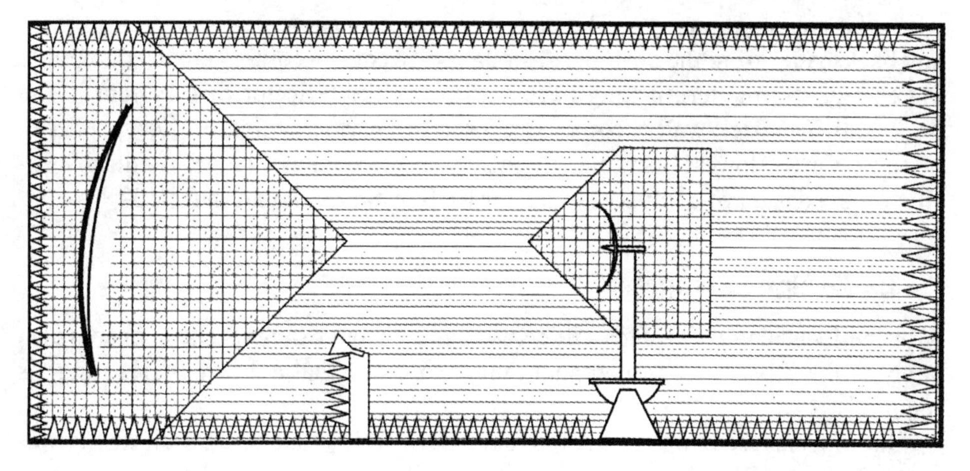

Figure 7.6 Alternate absorber layout for prime focus compact antenna range.

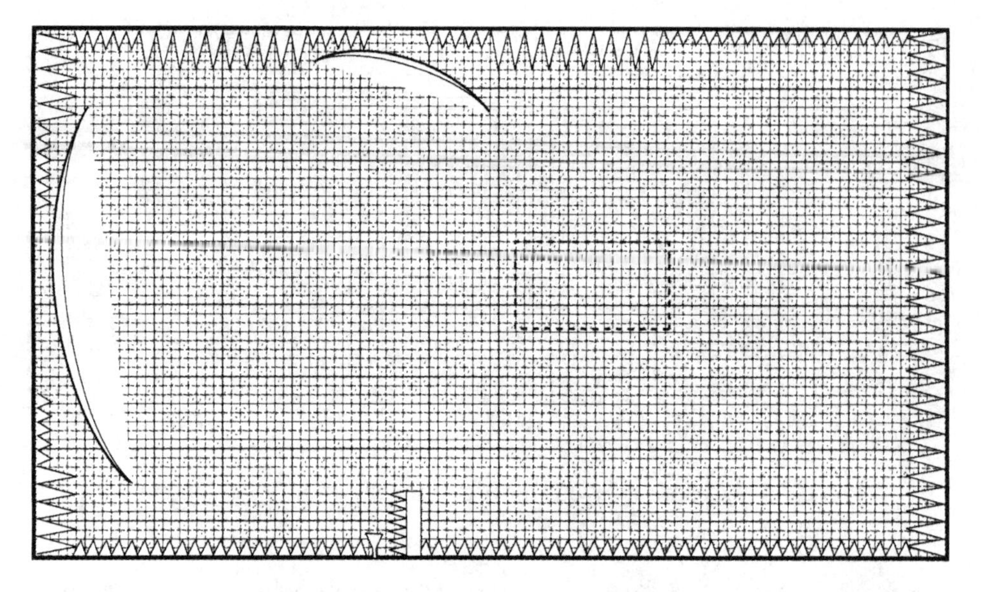

Figure 7.7 Absorber layout for dual reflector compact antenna range.

down in the 30 to 40-dB range. This is to allow an accurate measurement of the extraneous energy reflected from the chamber walls. The measured pattern sector looking toward the reflector is eliminated from the data reduction.

7.3 COMPACT RCS RANGES

7.3.1 Introduction

All of the above reflector types are used in the construction of compact RCS test facilities. The prime consideration in the chamber design is to reduce chamber background RCS. This is the energy reflected back to the instrumentation radar when the chamber is empty. The absorber layout must be designed to minimize any backscatter from the absorber installation. Wedge absorbers are used to guide the energy into the terminating back wall, from the reflector feed system to the back wall. The back-wall absorber is chosen to provide the least reflector energy possible for the lowest operating frequency of the compact range. A full discussion on the design of compact ranges for RCS testing in found in Ref. 3.

7.3.2 Design Example

A prime focus compact range will be used as a design example.The concepts also apply to the other types of compact ranges. Assume that a 2 to 18-GHz range is required.

The first surface to be considered is the back wall—that is, the wall that terminates the range. The absorber used on this wall is set by the lowest operating frequency of the reflector system. Assuming that the range is to operate down to 2 GHz, it is desirable that the least possible energy be reflected back toward the reflector system from this wall. The most that can be achieved is on the order of –50 dB at 2 GHz, using conventional materials. Using the normal incidence curve given in Chapter 3, the absorber needs to be on the order of 8 wavelengths thick. This means that the absorber must be on the order of 1.2 m (4 ft) thick, which is on the order of 73 wavelengths thick at 18 GHz. Beyond this point the tip scatter might become a problem. A tradeoff in the performance is required. Generally, 0.6 m (2 ft) of material is used for this frequency range on the back wall of compact ranges. At 2 GHz, the back-wall reflectivity is on the order of –44 dB, rising to greater than –50 dB at 18 GHz. When the chamber is empty and the compact range transmits a wave, the reflected energy from the back wall is reflected back to the reflector and the chamber background is limited by the returned energy. See Chapter 6 for a discussion on the residual backscatter from the back wall of a free-space RCS range. The same analysis applies to the compact RCS range. Experiments have demonstrated that if the back wall is slightly tilted, the reflected energy from this surface can be terminated on other surfaces of the chamber and not returned to the radar as a residual background level. It is recommended that the wall be tilted so that it is focused on the space between the top of the reflector and the ceiling of the chamber above the reflector. This is generally a tilt of only 5–8 degrees. Energy reaching the radar may be down as much as 30 dB, with respect to the original signal.

The side-wall absorber is chosen to be on the order of half the thickness of the end wall. Pyramidal absorber is used around the reflector system to terminate any stray energy from the feed system and from the edges of the reflector. A transition area is provided from the reflector area to the down range target area. This is illustrated in Figure 7.8. The remainder of the absorber is wedge material. Wedge ma-

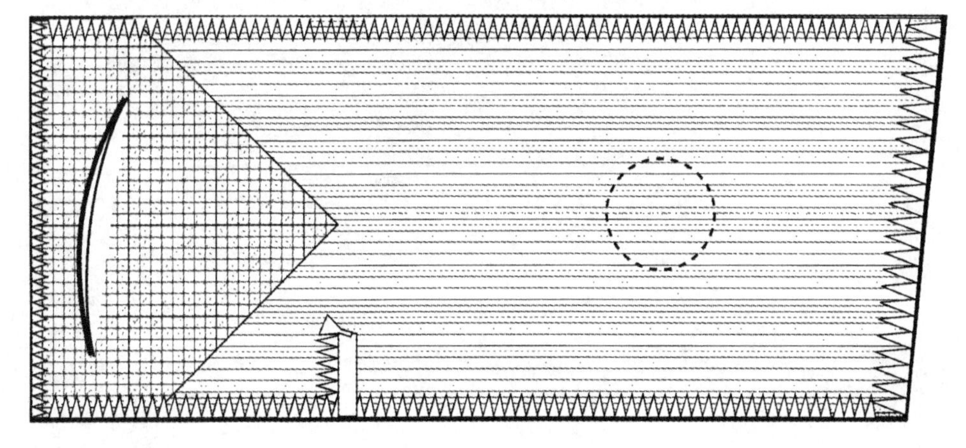

Figure 7.8 Illustrated is the absorber layout for prime focus compact radar cross-section ranges with a tilted back wall.

terial can be procured in long strips, 6–8 ft long. To minimize any physical discontinuities at the ends of the wedges, it is necessary to very carefully (very thin bond line) glue the rows end-to-end along the whole surface of the chamber right to the back wall. Each discontinuity is a source of backscatter that will limit the residual background of the chamber. It is also desirable that the impregnation of the wedges be as uniform as possible, because discontinuities in the loading of the absorber can also be a source of residual backscatter.

The absorber around the reflector is chosen to be the same size as the side-wall material. It is required to terminate the energy reflected off the edges of the reflector. The area immediately behind the reflector is often not covered in order to minimized cost.

7.3.3 Acceptance Testing

Acceptance testing of an anechoic chamber designed to house a compact RCS range is conducted after the instrumentation radar is installed and operating. The radar is used to develop images when the empty chamber is evaluated [4]. It is desirable that all discontinuities be minimized and the chamber residual RCS be on the order of a minimum of –70 dBsm. This can only be achieved with a tilt-wall chamber.

REFERENCES

1. R. Johnson, U.S. Patent 3,302,205, Issued January 1967.
2. V. J. Vokurka, U.S. Patent 4,208,661, *Antenna with Two Orthogonal Disposed Parabolic Cylindrical Reflectors,* Issued June 17, 1980.
3. E. F. Knott, *Radar Cross Section Measurements,* Van Nostrand Reinhold, New York, Chapter 8, 1993.
4. E. F. Knott et al., *Radar Cross Section,* 2nd edition, Artech House, Boston, p 543, 1993.

CHAPTER 8

INCORPORATING GEOMETRY IN ANECHOIC CHAMBER DESIGN

8.1 INTRODUCTION

One chamber design variable not used, except in a few installations, is shaping the geometry of the chamber. Shaping will take advantage of the properties of imaging, provide normal incidence illumination of the absorber for optimum performance, and permit use of transmission structures to perform electromagnetic measurements. This chapter covers (1) the tapered chamber that takes advantage of the method of images to control chamber reflections, (2) the double horn chamber which takes advantage of shaping to reduce chamber reflections, (3) the hardware-in-the-loop chamber with its special shaping, and finally, (4) the TEM device, which uses a shielded strip line geometry terminated in absorber to provide a compact testing system for EMC measurements.

8.2 THE TAPERED CHAMBER

8.2.1 Introduction

Tapered chambers were created in the early 1960s after it became obvious that rectangular chambers were not suited for VHF/UHF measurements. Emerson [1] determined that tapering one end of a chamber would cause the chamber to act like an indoor ground reflection range. Instead of trying to suppress the wall reflections, they were used to form a uniform illumination across the test region. The illumination difference between the rectangular and tapered chambers is il-

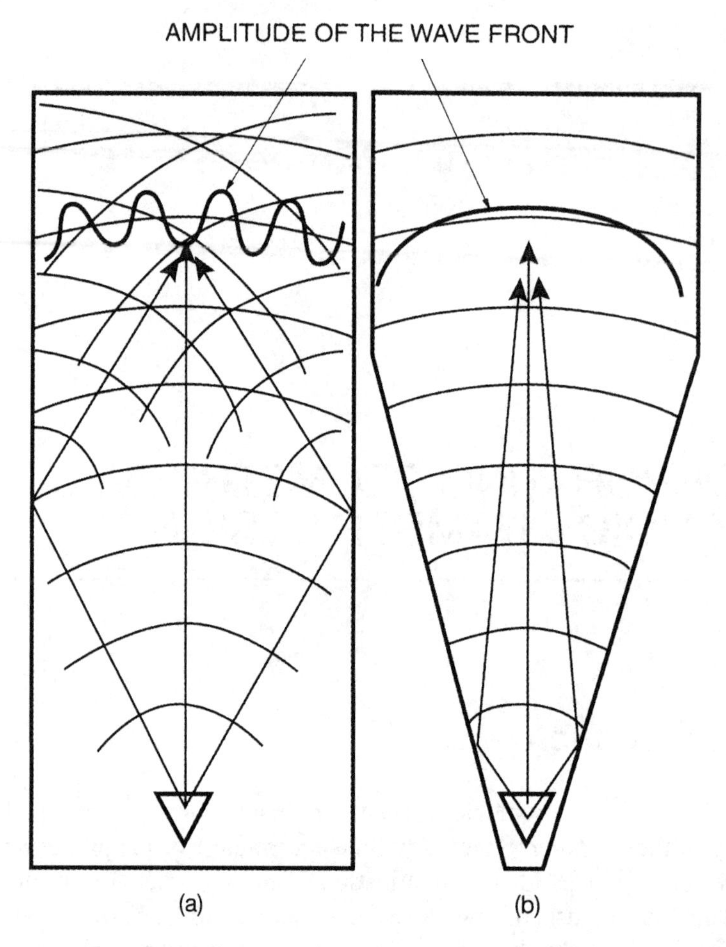

Figure 8.1 Test region illumination in (a) a rectangular chamber and (b) a tapered chamber.

lustrated in Figure 8.1. This is accomplished by forcing the wall images close together at the source end of the chamber. This effectively forms a source antenna array. The design was empirically derived and, even today, it is often necessary to tune new chambers to obtain acceptable test region performance. This was one of the first attempts at using chamber shaping to achieve an acceptable test environment for electromagnetic testing. The tapered chamber is capable of operating over very large frequency ranges, depending on how well the source antenna is incorporated into the tapered end of the chamber. The most common measurements performed in the chamber are to characterize low-frequency antennas. Some tapered chambers are used for low-frequency satellite testing, whereas others are used for low-frequency RCS testing. The free-space range equation applies to these chambers; thus the size of the test aperture sets the range length. The chambers are expensive to build, and the largest known chambers are on the order of 55 m (175 ft) in length. The cost is less than comparable-length rectangular chambers, because the bulk of the materials used are low in cost compared to the thicker materials required to achieve equivalent performance versus frequency in the rectangular chambers. A comparison of rectangular and tapered chamber performance is shown in Figure 8.2.

A comparison of various performance criteria for the rectangular and tapered chamber is given in Table 8.1.

8.2.2 Antenna Testing

8.2.2.1 Design Considerations. The following factors need to be considered in the design of this type of chamber:

a. Frequency of operation
b. The level of extraneous energy permitted versus the sidelobe levels of the antenna under test.
c. The size of the antenna under test.
d. The space available to construct the chamber.
e. Shielding requirements.

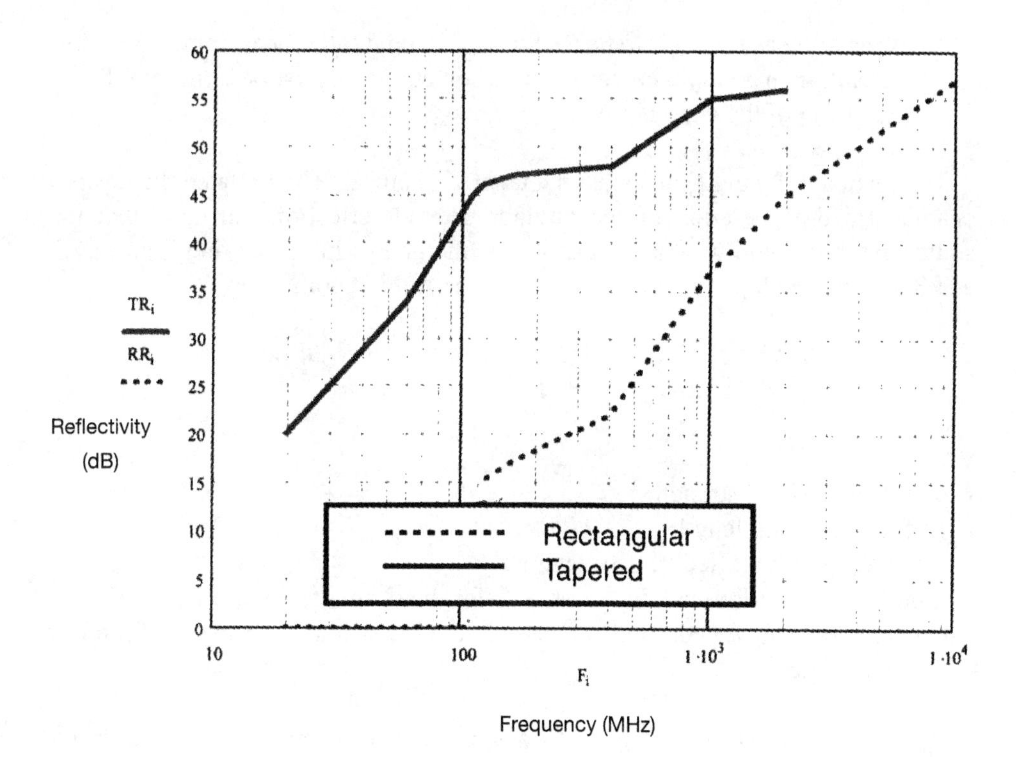

Figure 8.2 Reflectivity levels measured in large rectangular and tapered chambers.

Table 8.1 A Comparison of Tapered Versus Rectangular Electromagnetic Anechoic Chambers

Criteria	Tapered	Rectangular
1. Antenna patterns	Excellent low, mid-, and high frequencies[a]	Poor low frequency, good mid-frequency, excellent high frequency
2. Radar cross section	Monostatic only	Monostatic and bistatic
3. Frequency sensitivity	High	Low
4. Axial ratio	≤0.5 dB	≤0.1 dB
5. Cross-polarization	≤25 dB	≥35 dB
6. Source antenna choice	≤15 dB	Limited only by far-field criteria
7. Amplitude taper (TR)	Frequency-dependent	Frequency-dependent
8. Phase deviation (TR)	Frequency-dependent	Frequency-dependent
9. Boresite error	Potentially high	Low
10. Swept frequency measurements	Must be carefully calibrated	Ideal configuration
11. Cost factor (same range length)	1.0	1.5 to 2.0
12. Free-space loss	No	Yes

[a]Very careful construction of the conical tip is required to achieve high-frequency performance in a tapered chamber. A pseudorectangular mode is often used for high-frequency operation.

f. Types of measurements to be performed. If axial ratio measurements are important, then a very well constructed conical horn must be built into the tapered end of the chamber.

Hickman and Lyon [2] analyzed the tapered chamber. They related the tapered chamber to the operation of the outdoor ground reflection range, in that the source antenna could be analyzed using the method of images. From Figures 8.3 and 8.4, they developed the relationship for the field at point P as

$$E(P) = E_D[e^{-j2\pi R_D/\lambda} + C_1(R_D/R_{R1}e^{-j2\pi R_{R1}/\lambda} + R_D/R_{R2}e^{-j2\pi R_{R2}/\lambda}) + 2C_2R_D/R_o e^{-j2\pi R_o/\lambda}] \quad (8\text{-}1)$$

where
E_D = the direct field strength,
R_D = the direct path length,
R_{R1} and R_{R2} are the reflected path lengths,
C_1 and C_2 represent reflection coefficients for the chamber side walls.
The phases for the two constants are assumed to be π radians because of the low grazing angles.

The magnitude of the constants C_1 and C_2 were determined from data developed by King et al. [3] using analysis developed by Jordan [4], for the reflection coefficient as a function of the angle of incidence and polarization. The result of

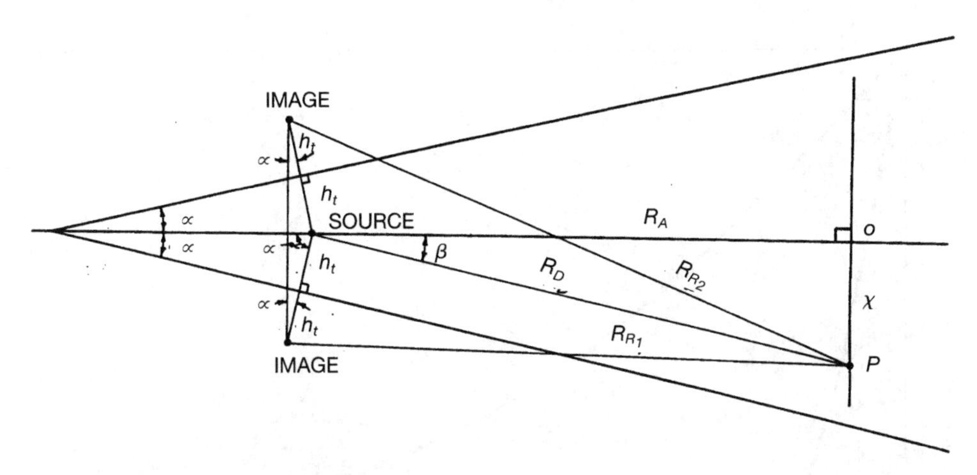

Figure 8.3 Simplified plane geometry of the tapered section of the chamber showing quantities used in the calculation.

this analysis is plotted in Figure 8.5. Using these constants, and assuming the tapered chamber geometry shown in Figure 8.6, equation (8-1) was solved and plotted in Figure 8.7. The chamber was then field probed to determine the experimental field taper in the test region, the results of which are shown in Figure 8.8 for both polarizations. The difference in the taper between the calculated and the measured field probes is that the pattern of the source antennas was neglected in

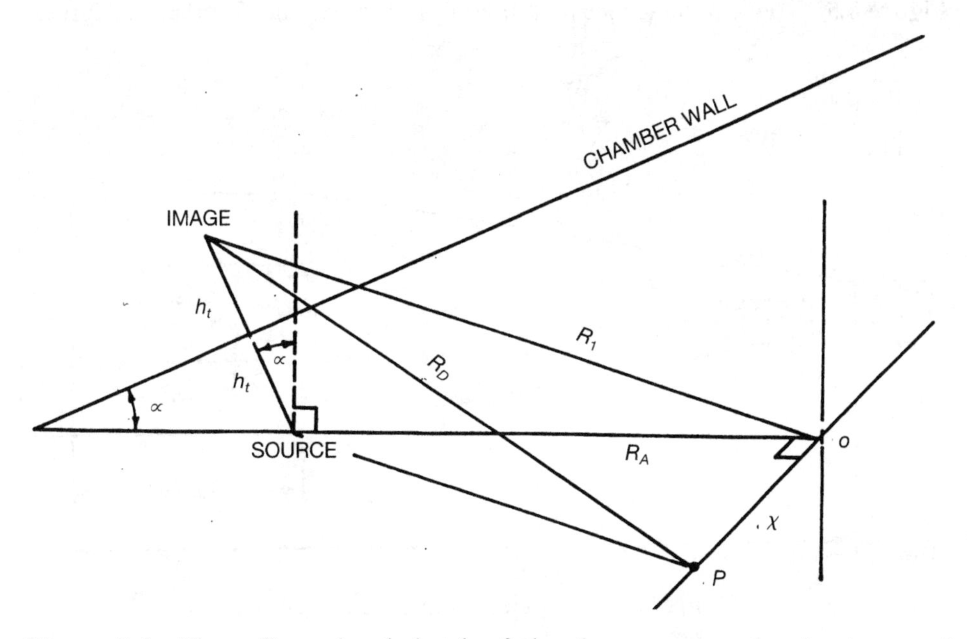

Figure 8.4 Three-dimensional sketch of chamber geometry showing image of reflection from upper wall to field point on the horizontal plane.

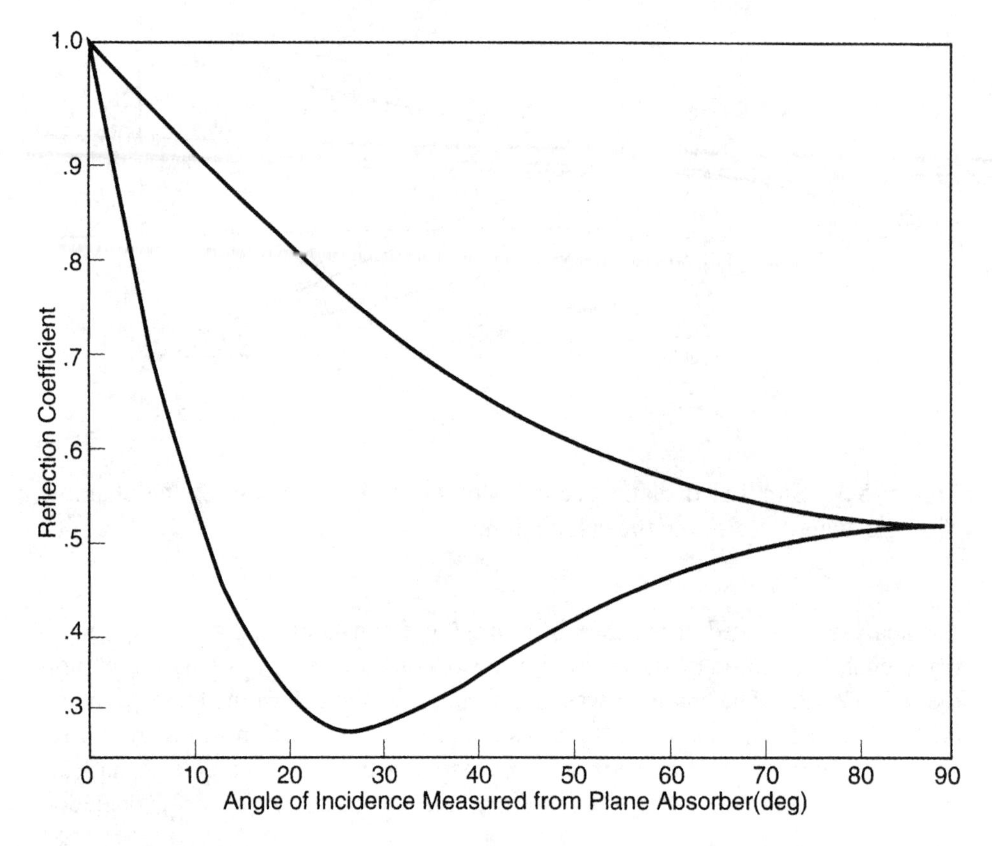

Figure 8.5 Angle of incidence measured from plane of the absorber in degrees.

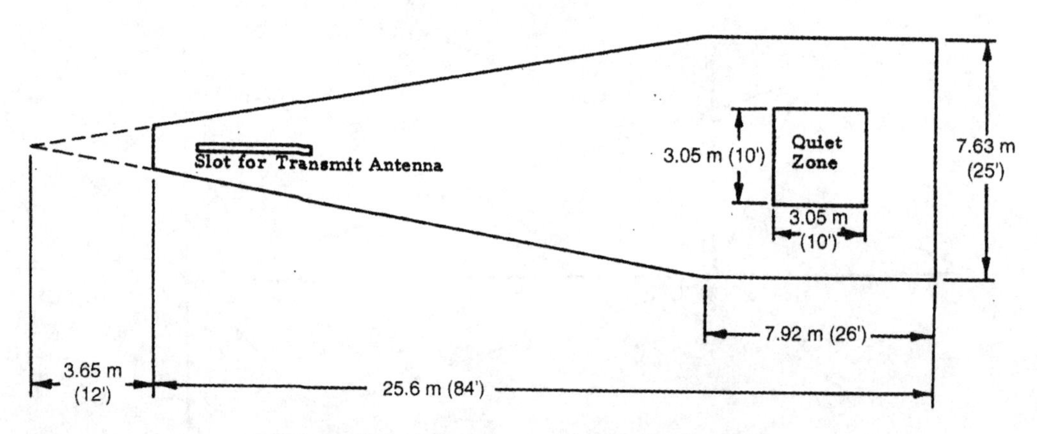

Figure 8.6 Plan view of anechoic chamber.

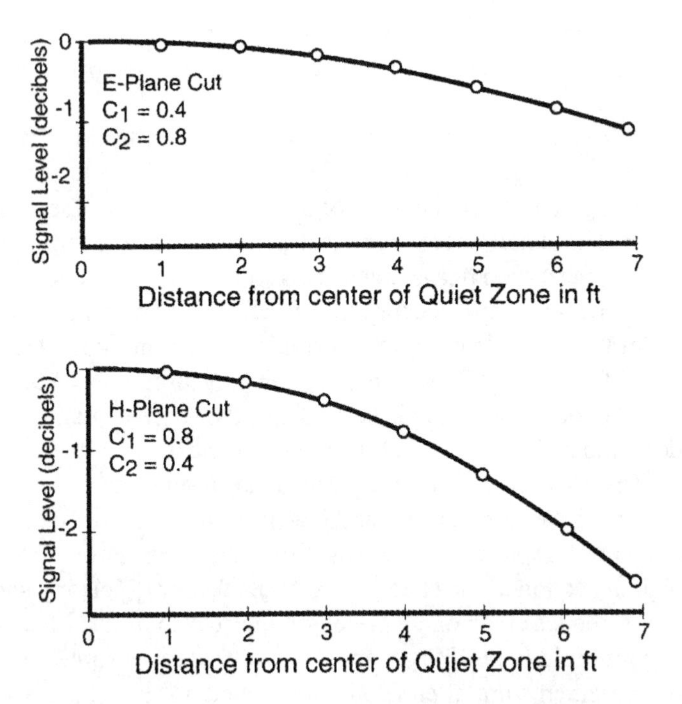

Figure 8.7 Calculated amplitude taper across the test region in both E and H-planes of the transmit antenna as a function of distance from the center of the test region.

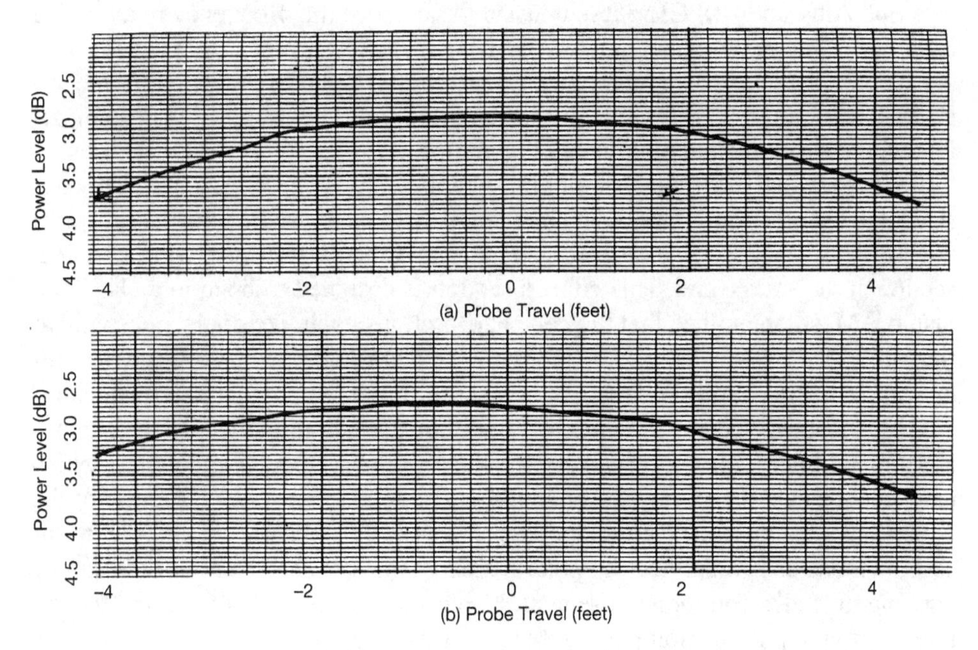

Figure 8.8 Transverse field probe cuts representing excursions of ±4.5 ft across the test region. (a) E-plane cut, (b) H-plane cut.

the calculations. The image antenna magnitudes are attenuated, thus broadening the pattern of the source array and providing a slower roll-off in the measured field versus the calculated.

The low-end frequency determines the thickness of the backwall absorber. The size of the desired test region determines the width/height. The test region for 1-dB amplitude taper is generally on the order of one-third of the cross section of the rectangular section. The actual test aperture is dependent upon the range equation, because this sets the permitted phase taper across the test region. The taper angle depends on the frequency range, because the effective antenna array spacing formed by the taper must be very tight to prevent ripples appearing in the test region illumination. Consider Figure 8.3 , which shows an elevation view of the tapered end of a chamber. The location of the phase center of the source antenna within the tapered end determines the corresponding images in the walls of the taper, as shown in the figure. These can be considered a ring array around the center antenna. The array spacing and the amount of loss in the walls of the chamber determine the effective array factor. The pattern of the array forms the illumination to the test region in the rectangular section of the chamber. Antenna theory [5] clearly shows that the array factor must be less than one wavelength in order to prevent the array pattern from forming multiple lobes. However, this requirement is somewhat softened if the outer array elements are attenuated with respect to the center array element, which is the case in the tapered chamber with its lossy walls. The effect of the loss in the walls are to broadband the operation of the chamber, as is found in any lossy network solution. Experience has shown that the included angle should be less than 36 degrees for frequency coverage up to 6 GHz and less than 30 degrees for chambers operating up to 18 GHz. It is unusual to go above this frequency in a tapered chamber. Sometimes, a dual mode is used where the low-frequency operation uses the taper, and a semirectangular mode is used for the higher frequencies. In the latter case, the patches in the walls, ceiling, and floor must be designed to be compatible with the semirectangular mode of operation.

Because the walls of the chamber must be metal, it is common practice to construct the chamber out of modular shielding panels. Some chambers are built with the panels arranged into a tapered geometry and then terminated with a conical section at the source end. This conical section is designed to open in a clamshell fashion. Most are built so that the cone half-sections open like doors and swing to the right or left. A few have been built with the clamshell opening upward. Another approach is to build the tapered chamber within a rectangular shielded chamber. The tapered section is constructed within the shield using plywood and a sheet metal liner. This approach uses standard modular shield components, which can keep the cost down. It is common to build a permanent stand at the end of the taper to make it possible for the operator to access the cone section in order to insert antennas and to service the polarization positioner used to rotate the source antenna. It is also common to use a telescopic slide arrangement to mount the antenna onto the polarization positioner using large aluminum tubing. This permits locating the antenna within the conic section at the optimum position for proper operation of the tapered chamber. Prior to conducting the chamber acceptance procedures, the antenna positions versus frequency and polarization are determined by probing the test region for optimum operation.

Gain measurements must be conducted by comparison to a standard gain antenna. The site attenuation of a tapered chamber does not follow the Friis free-space transmission formula, as does a rectangular chamber.

8.2.2.2 Design Examples

(a) Conventional Tapered Chamber. Assume that we wish to conduct measurements from 500 MHz to 18 GHz. At 500 MHz we wish the chamber to have a reflectivity level of –35 dB. The desired performance is to increase at a 6 dB per octave rate up to 18 GHz. The test region is to be 1.2 m (4 ft) in diameter. What does our chamber design need to be? The chamber cross section must be 3.6 m (12 ft) in order to provide at least a 1-dB amplitude taper across the test region. Referring to our absorber performance data in Figure 8.9, the back-wall absorber needs to be approximately 0.91 m (3 ft) thick to achieve the –35-dB requirements. Because this is on the linear portion of the performance curve, the chamber performance will increase at a 6 dB per octave rate with frequency. Because we wish to operate up to 18 GHz, set the cone angle at no more than 30 degrees. The ta-

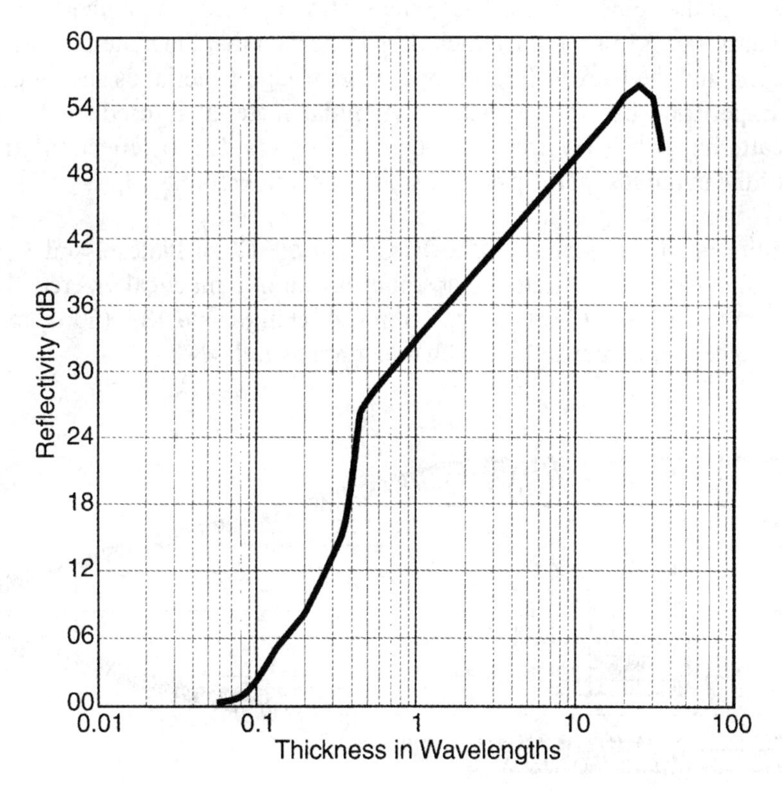

Figure 8.9 Normal incidence performance of pyramidal absorber.

pered section of the chamber then needs to be on the order of 6.7 m (22 ft) long. The tip of the cone is truncated so that the source antenna end of the chamber is about 0.30 m (1 ft) in diameter. Thus, the actual overall taper length will be 6.25 m (20.5 ft). It is standard practice to make the conical section long enough to clear the largest antenna to be mounted. This will be a horn antenna with a diagonal dimension of 0.30 m (1 ft), and because the absorber inside the cone is 0.15 m (0.5 ft) thick, the cross section must be 0.61 m (2 ft) total. The minimum length for the clamshell is 0.61–0.91 m (2–3 ft). The conic section needs to be twice this, or 1.8 m (6 ft). The transition from round to square must be very gradual so that a smooth surface is provided to minimize discontinuities. It is recommended that the square-to-round transition be on the order of 2.4 m (8 ft). The remainder of the taper will be a truncated square pyramid with a length of 2.6 m (8.5 ft). This is joined to a rectangular section. The rectangular section is 4.6 m (15 ft) long: 0.9 m (3 ft) for the absorber and 3.7 m (12 ft) for the test region. Thus, the chamber overall length is 10.8 m (35.5 ft). The side walls in the test region are covered with absorber that is one-half to one-fourth the thickness of the back wall. It is recommended that 0.46 m (1.5 ft) of pyramidal material be used in the test region patch located opposite the 1.2-m (4-ft) test region. The chamber design is illustrated in Figure 8.10. Wedge material is used in all other areas of the chamber, starting with 0.305 m (1 ft) material in the rectangular section and tapering down to 0.15 m (0.5 ft) at the conic section. The conic section is lined with a plug of uniform-loaded absorber 0.15 m (0.5 ft) thick. It is important that the material between the test region and the back wall be composed of wedge material as shown in Figure 8.11. Experience has shown that if pyramidal material is used in this region, backscattering will occur from the chamber corners due to reflections from the sides of the pyramids on the side walls, as shown in Figure 8.12.

(b) Example of a Practical Tapered Chamber. Color Plates 2 and 3 (located in the color section) illustrate the absorber layout in a practical tapered chamber. The chamber is 7.3 m (24 ft) wide, 7.3 m (24 ft) high, and 18.3 (60 ft) long. The performance requirements for this chamber are as follows:

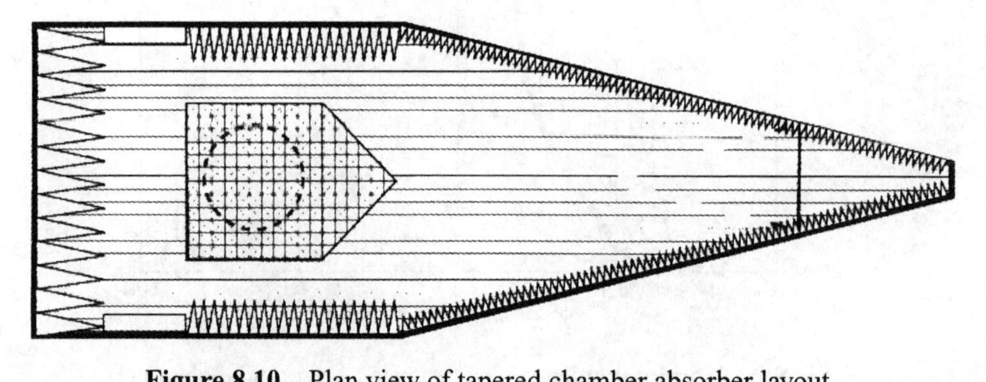

Figure 8.10 Plan view of tapered chamber absorber layout.

Figure 8.11 Elevation view of tapered chamber absorber side-wall layout. (Photograph courtesy of Lehman Chambers, Inc., Chambersburg, PA.)

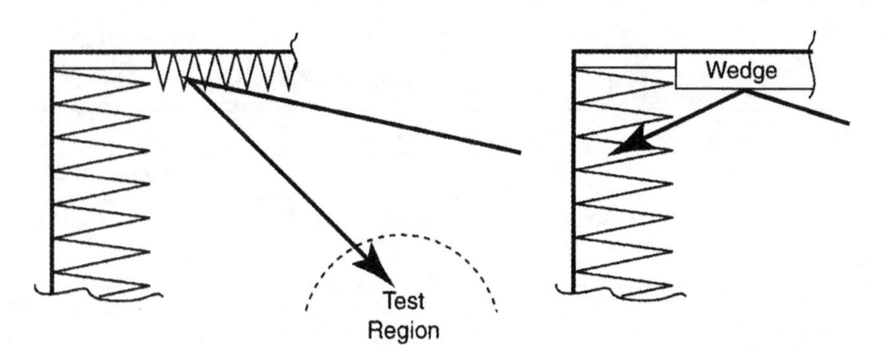

Figure 8.12 Reflections that can occur if pyramidal material is used in the area near the back wall of a tapered chamber.

Frequency (GHz)	Specified Reflectivity (dB)	Expected Reflectivity (dB)
0.100	–22	–25
0.200	–27	–30
0.400	–35	–38
4.000	–45	–50

The layout using the procedures outlined above resulted in an absorber layout as follows:

The back wall was selected to be a 1.2-m (4-ft) pyramidal absorber with a picture frame installation. The side-wall patches were 0.45 m (1.5 ft) pyramidal, oriented as shown in the photographs. The wedge material in the rectangular section of the chamber was selected to be 0.3 m (1 ft) thick. The absorber in the tapered section started at 0.3 m (1 ft) and tapered to 0.15 m (0.5 ft) thick at the cone. The cone material is a 0.15-m (0.5-ft)-thick solid absorber. The conic section was designed to swing out and was designed to have an antenna cantilevered from a polarization positioner located at the end of the taper. The measured performance of the chamber is given in Table 8.2.

(c) Automatic Feed System. An example of how an automatic feed system can be incorporated into a tapered chamber is illustrated in Figure 8.13. The conic section is clamshell hinged such that, when opened, the feed horns can rotate into the correct axial location in the conic section to provide the proper field illumination in the test region.

(d) Portable Tapered Chamber. A very useful form of the tapered chamber created by Holloway [6] is shown in Figure 8.14. His chamber consists of a conic section terminating into a cylindrical section. The overall length is 4.6 m (15 ft), and diameter is 1.5 m (5 ft) It is mounted on a frame supported by casters. The end wall is terminated in a 0.46-m (18-in.) absorber, making the chamber useful down to about 300 MHz. The tapered section is lined with absorber that blends

Table 8.2 Tapered Chamber Performance

	Reflectivity (dB)							
	Typical				Worst Case			
	Transverse		Longitudinal		Transverse		Longitudinal	
Frequency (GHz)	H	V	H	V	H	V	H	V
0.100	–27	–30	–32	–21	–25	–25	–27	–21
0.200	–45	–40	–36	–35	–37	–31	–31	–31
0.400	–46	–45	–37	–43	–42	–42	–33	–40
4.000	–51	–52	–57	–58	–50	–51	–56	–53

Note: Some interaction occurred with the antenna positioner.

Figure 8.13 An example of an automatic feed system that can be used in tapered chambers. [Photograph courtesy of EMC Test Systems (ETS), Austin, TX.]

into the 0.3-m (1-ft) pyramidal absorber in the test region. The test region is about 0.46 m (18 in.) in diameter. The axial ratio of this chamber is less than 0.1 dB from 2 to 18 GHz. It is an excellent chamber for small antenna development or product testing.

(e) Chebyshev Absorber Terminated Tapered Chamber. In Ref. 7, a chamber using a Chebyshev pyramidal absorber was investigated. The chamber has a special low-frequency antenna [8], which fits into a 20-degree cone angle. This chamber is illustrated in Figure 8.15. The back wall consists of Chebyshev material that is 0.81 m (32 in.) thick, but has the performance of a conventional 1.83-m (6-ft) absorber and is on the order of –40 dB at 400 MHz [9]. Even with the 20-degree cone angle, the overall chamber length is held to less than 8.5 m (28 ft) The chamber cross section is 2.74 m (9 ft). The test region can be on the order of 0.9 m (3 ft) for a 1-dB taper. The actual test aperture size is a function of the range equation that sets the phase taper across the test region.

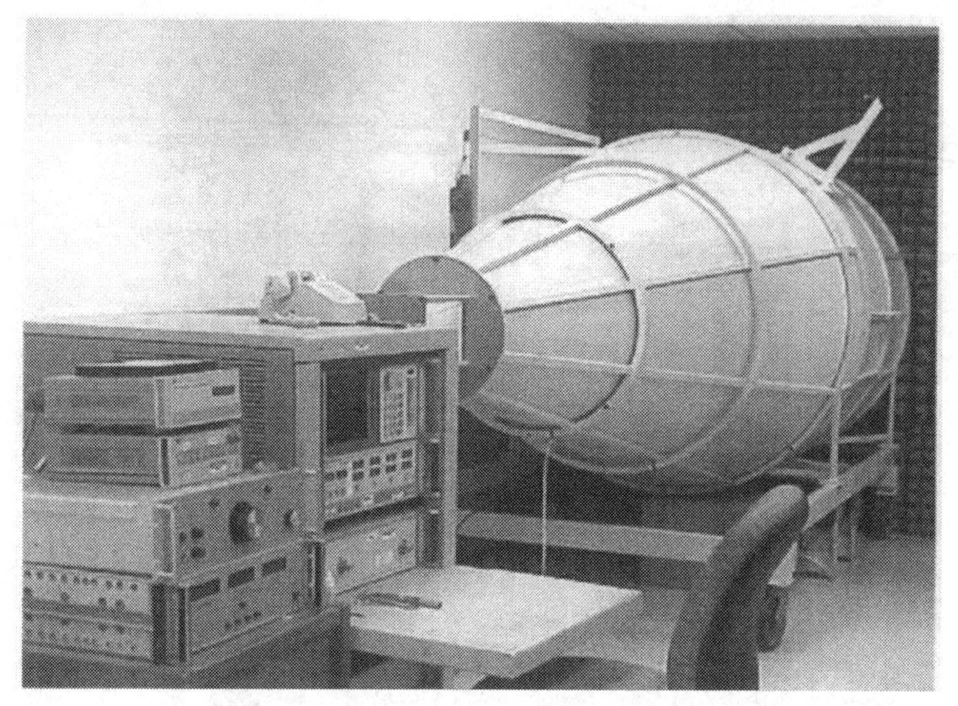

Figure 8.14 Portable tapered chamber used for electrically small antennas. (Photograph courtesy of the Boeing Company, Mesa, AZ.)

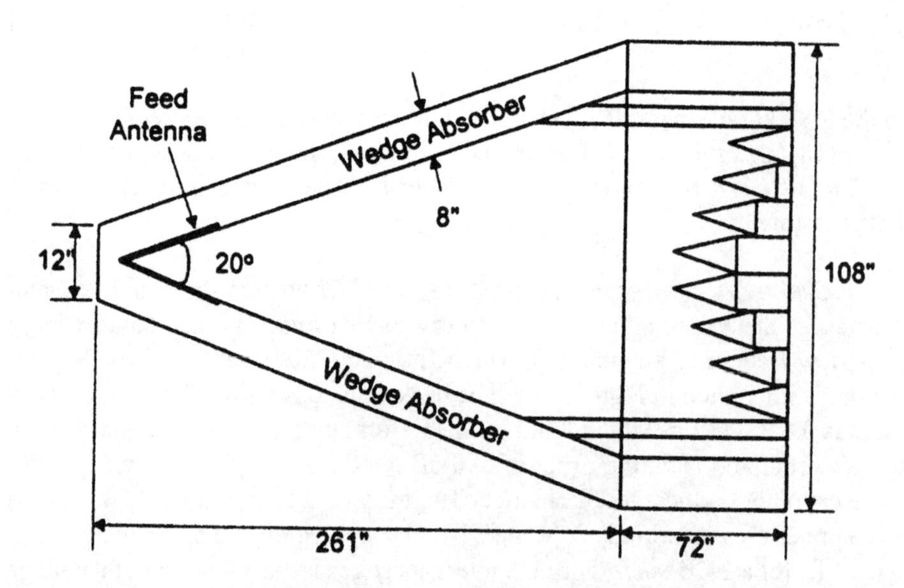

Figure 8.15 Illustrated is a small tapered chamber with a Chebyshev pyramidal absorber.

8.2.2.3 Acceptance Test Procedures.

The Free-Space VSWR Method is the standard procedure used for acceptance testing. It is recommended that at least five frequencies be tested for both vertical and horizontal polarization. If a horizontal probe is used, it should be used at the top, middle, and bottom of the test region. The procedure is detailed in Section 9.3.2.

If a high-performance chamber is required, then the chamber should be characterized by using the planar scanner method described in Section 9.3.4. This technique provides a clearer picture of the sources of extraneous energy and permits the chamber performance to be optimized.

8.2.3 Radar Cross-Section Measurements

8.2.3.1 Design Considerations.

When a tapered chamber is designed for radar cross-section measurements instead of antenna pattern measurements, the only extra consideration is the distance between the test region and the back wall. When hardware gating is employed, it is necessary to have sufficient distance between the test region and the back wall so that the radar can gate out the reflections from the wall. This is a matter of radar design and instrumentation considerations. The residual chamber backscatter is estimated using the procedures developed in Section 6.3.1.

8.2.3.2 Acceptance Test Procedures.

The same procedures used in a free-space rectangular chamber are also used in tapered chambers as described in Section 9.3.5.

8.3 THE DOUBLE HORN CHAMBER

8.3.1 Introduction

The double horn chamber [10, 11] was developed to improve low-frequency performance of the rectangular chamber when a ground plane was required. As discussed in Chapter 5, the reflectivity of microwave absorber drops off sharply as the angle of incidence increases, such as in the case of an absorber installed on the side walls of an anechoic chamber. This performance drop is very fast when the absorber thickness is small in terms of wavelength. The double horn design uses chamber shaping to guide and force the absorber incidence angles to be more nearly normal to the incoming wave front, thus obtaining higher reflection loss. These principles are illustrated in Figure 8.16. Shaping of the chamber is accomplished using ray-tracing techniques to minimize the amount of single bounce energy reaching the test region of the chamber. This is illustrated in Figure 8.17.

8.3.2 Antenna Testing

8.3.2.1 Introduction.

Usually, the rectangular chamber is limited to operations above 1 GHz. It is not economical to go below this frequency because the

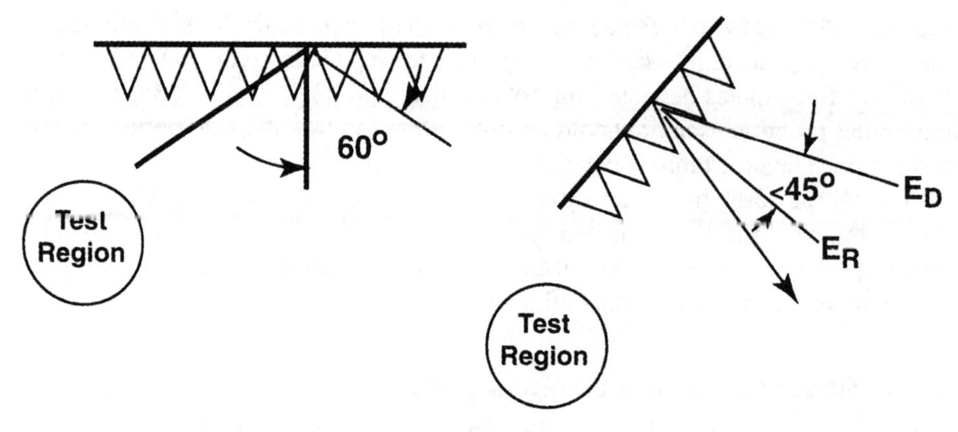

Figure 8.16 The principles used in the design of the double horn chamber is illustrated.

side-wall absorber has to be very thick in terms of wavelength in order to provide a sufficient amount of bistatic loss in order to reduce the side-wall reflections to a sufficient level. This limitation can be overcome by shaping the chamber so that the angle of incidence on the chamber walls is more nearly within 45 degrees, the angle at which pyramidal absorber performance begins to drop off.

8.3.2.2 Design Example. Low-frequency antenna testing (i.e., testing below 1 GHz) has recently become more common because of the growth of wireless equipment. A common test requirement is antenna patterns of equipment that operates in the 890-MHz band. Assume that it is required to test a hand-held wireless unit with a low-gain antenna to a repeatability of better than ±0.25 dB. This means that reflected energy reaching the test region must be less than –31 dB. Further assume that the test region is on the order of 0.6 m (2 ft) in diameter. The minimum range length must meet the requirements of the range equation up

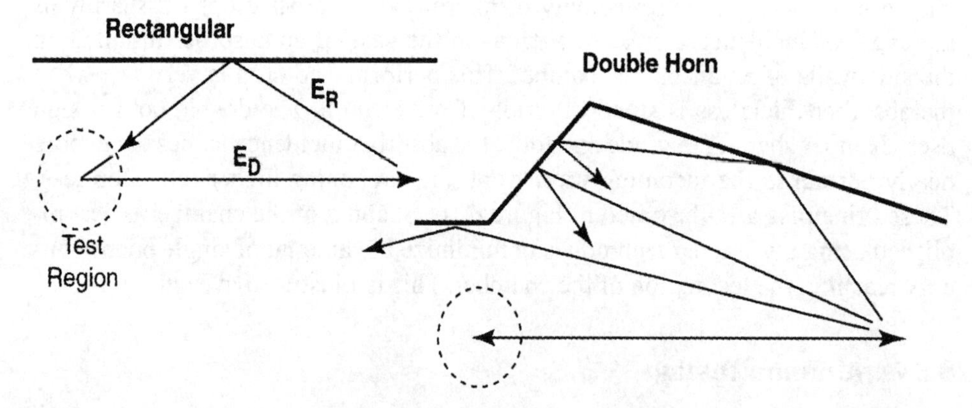

Figure 8.17 How shaping is used in the design of the double horn chamber.

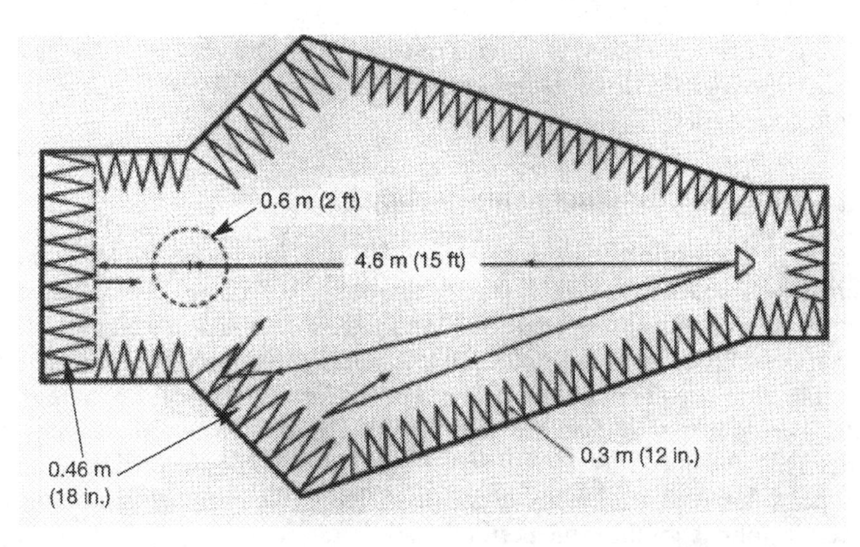

Figure 8.18 The design of a low-frequency microwave chamber using the double horn chamber.

through 1.8 GHz. Using the range equation, the range length needs to be a minimum of 4.6 m (15 ft). Laying this out on grid paper, the chamber appears as is illustrated in Figure 8.18. Note that the rays from the source antenna via the chamber walls do not reach the test region by a single bounce, except those from the wall terminating the chamber. This wall sets chamber performance. The –31-dB requirement at 890 MHz sets the absorber reflectivity at normal incidence. Thus, the terminating wall needs to be 0.46 m (1.5 ft) thick. The receiving tapered walls are lined with a 0.46-m (1.5-ft) pyramidal absorber, and the transmitting tapered wall is lined with 0.3 m (1 ft) of wedge material. Two potential sources of extraneous energy are located near the transitions from the rectangular ends into the tapers, and they will require special treatment during the absorber installation. On the transmit end, Chevechev wedge absorber can be used with good effect to reduce the extraneous signal levels by approximately 8 dB better than with a standard 0.3-m (1-ft) absorber.

The expected performance for the above design is as follows:

Frequency (MHz)	Reflectivity (dB)
890	–31
1000	–32
1800	–35
3000	–40

8.3.2.3 Acceptance Testing. The best method of acceptance testing for a chamber used for microwave antenna measurements is the Free-Space VSWR

Method described in Section 9.3.2. It is recommended that the procedure be conducted at three heights in the test region at a minimum of five test frequencies and for both polarizations.

8.3.3 Emissions and Immunity Testing

8.3.3.1 Introduction. The double horn chamber is uniquely suitable for performing low frequency testing in the 30- to 1000-MHz region. The design saves chamber volume and surface area when compared to a similar rectangular chamber. For example, as much as 30 percent less surface area is required for a shielded double horn chamber compared to a rectangular chamber of the same range length. The double horn chamber described in Section 10.4.3 demonstrates this saving. That chamber was constructed using conventional dielectric absorbers. With the widespread availability of ferrite material, the chamber can be further reduced in volume for the same performance.

8.3.3.2 Design Considerations. Assume that a chamber with a 3-m test region is to be used for 3- and 10-m measurements. Industry has developed a set of rectangular designs which are on the order of 18.9 m (62 ft) long, 11.6 m (38 ft)

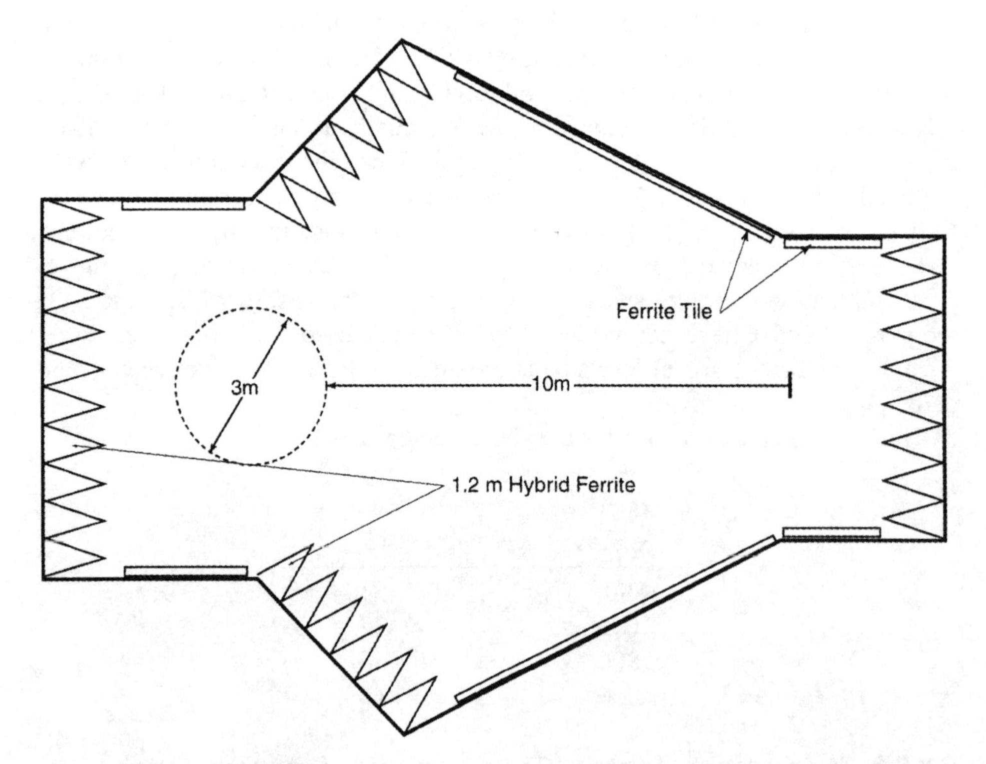

Figure 8.19 Plan view of an EMC chamber using the double horn design.

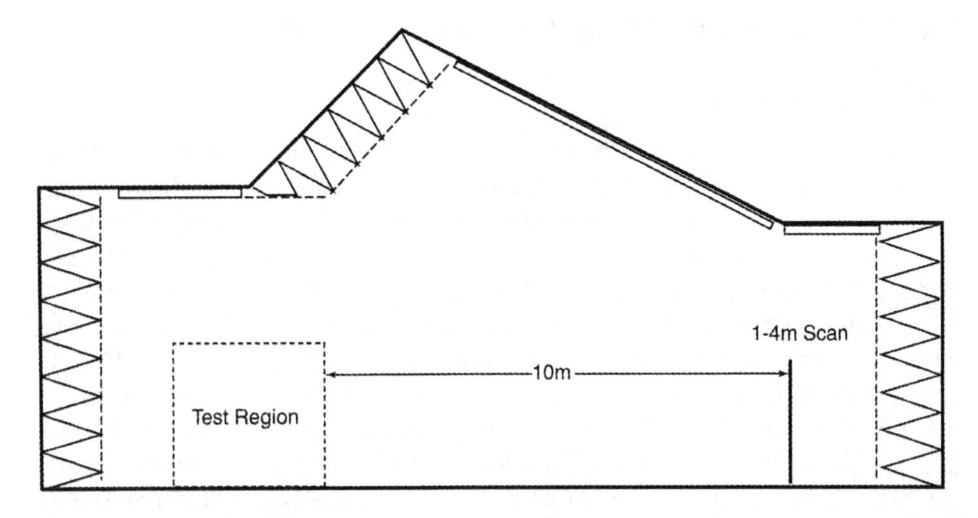

Figure 8.20 Elevation view of an EMC chamber using the double horn design.

wide, and 8.5 m (28 ft) high. An equivalent double horn design would be on the order of 18.3 m (60 ft) long, nominal width of 7.3 m (24 ft) with maximum width of 13.4 m (44 ft) and an average height of 7.3 m (24 ft), as shown in Figures 8.19 and 8.20. The actual surface area difference between the rectangular and double horn chambers is 893 – 668 or 225 m^2 (10,008 – 7190, or 2818 ft^2). This is a 26%/28% savings in absorber/shielding. This is slightly offset by the additional cost in the design process due to the compound angles in the building construction. Additional savings are possible, because the high-performance material need only be used on each end wall and on the tapered surface adjacent to the test region. The other surfaces can be treated with lower-performance material.

When the chamber is used for emission testing, the chamber floor is left bare, and the range is set up with an antenna located 10 m from the center of the test region. This antenna scans from 1 to 4 m in height as the measurement is conducted. The site attenuation for the range shall be less than 4 dB different than that obtainable with an open-area test site. Based upon experience in the industry, this can be held to less than 3 dB.

Immunity testing requires that a uniform field illuminate the device under test at a constant field strength, which ranges from 3 to 200 V/m, depending on the test requirements. The most common arrangement is to set the source 3 m from the test aperture and fix its height at 1.5 m. To eliminate reflections from the floor, an area between the source and the test aperture is covered with absorber.

8.3.3.3 Acceptance Test Procedures. The standard procedures for alternate test facilities called out in ANSI C63.4 and in EN 61000-4-3 are used to validate the performance of the double horn chamber.

8.4 THE MISSILE HARDWARE-IN-THE-LOOP CHAMBER

8.4.1 Introduction

Hardware-in-the-loop missile test systems provide the ability to simulate real-time missile flight scenarios from prelaunch to target intercept. Hardware-in-the-loop simulation has been conducted since the early 1960s on a variety of air-to-air and ground-to-air missiles. The systems are housed in specially designed anechoic test chambers. They consist of actual missile hardware, a flight motion table, computer-controlled IR and RF sources utilized to simulate maneuvering targets, and a real-time six-degree-of-freedom computer simulation. The RF sources generally consist of an array of RF horns positioned on a wall opposite the missile seeker. By radiating emissions, this target presentation system can provide the ability to simulate multiple targets, electronic countermeasures (ECM), and background clutter. ECM capabilities can be provided to simulate integrated multipurpose jammers or actual flight-test jamming pods.

8.4.2 Design Considerations

These chambers are normally pie-shaped in the azimuth plane, with the missile motion table located at the tip of the pie. The array of antennas is mounted in the

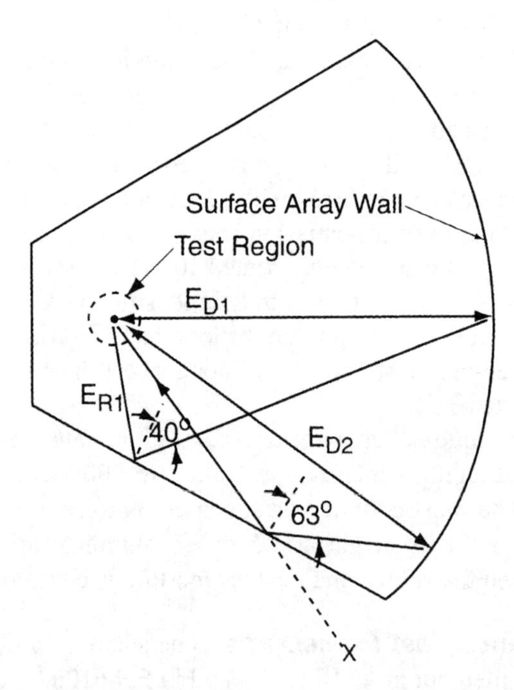

Figure 8.21 How ray tracing is used in the design of a hardware-in-the-loop anechoic chamber.

circular-shaped wall terminating the chamber. The anechoic design is a variation of the procedures used to design rectangular chambers. Antenna locations at the top, bottom, and at the azimuth extremes are selected as source antennas.The anechoic design is developed so that the extraneous signal level of the missile seeker is within the specified requirements. An example of the ray-tracing layout is shown in Figure 8.21. The antennas are boresighted to the missile seeker antenna mounted in the motion simulator. Thus, allowance is made for this in the angle of the energy radiated from the source antennas when computing the level energy illuminating the surfaces of the chamber.

8.4.3 Design Example

A typical layout is shown in Figure 8.22, which is a picture of the hardware-in-the-loop chamber located at the Naval Research Laboratories in Washington, D.C. Note the antenna array on the curved back wall.

Another target motion simulator is pictured in Figure 8.23. This is a dual antenna system where the target antennas are on two independent X–Y positioners on a common structure. Both targets can be tracked simultaneously. Figure 8.24 is a

Figure 8.22 The array wall in a large hardware-in-the-loop chamber at the Naval Research Laboratories. (Photograph courtesy of NRL, Washington, D.C.)

Figure 8.23 Illustration of two-target motion simulation (Used with permission.)

photograph of such a system, as seen from the focal point of the range.

8.4.4 Acceptance Test Procedures

The Free-Space VSWR Test Method is used to evaluate chamber performance. Several of the horns from the array wall are used as source antennas to ensure that performance requirements are met as the various antennas are excited during a test. The evaluation is normally limited to ±90 degrees of boresight due to the physical limitation imposed by the flight simulators used to hold the missiles.

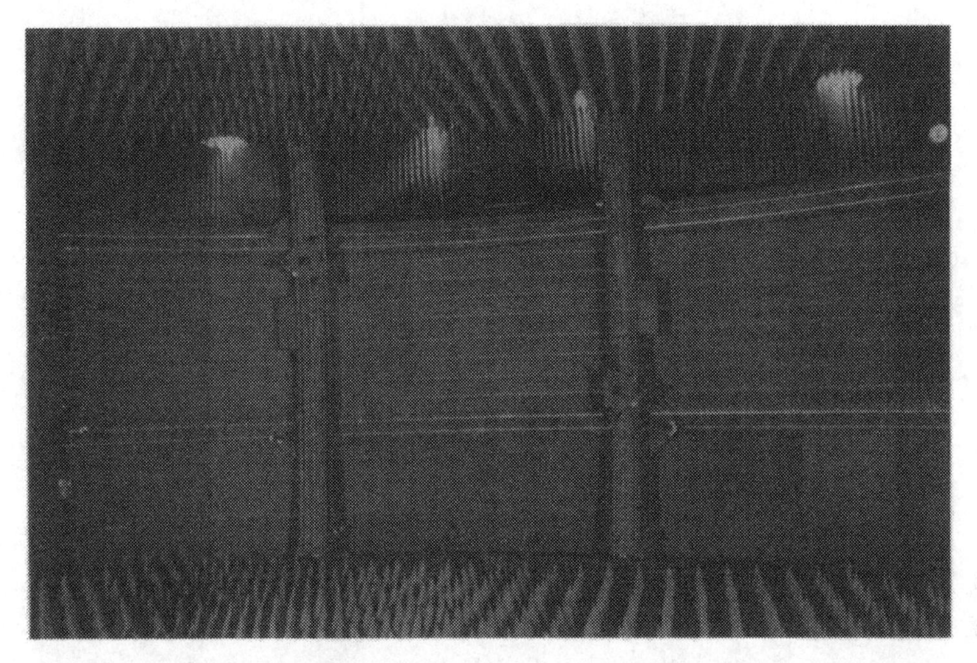

Figure 8.24 View down range in a two-target motion simulation chamber. (Used with permission of Carco Electronics, Inc.)

8.5 CONSOLIDATED FACILITIES

8.5.1 Introduction

Consolidated facilities are those that combine two or more range concepts into a common anechoic chamber. A combination of a free-space range for low-frequency antennas with a compact range for high-frequency antennas is possible by sharing the same positioning and range instrumentation. This is accomplished by arranging the chamber into an "L" shape, where one leg houses the compact range reflector, and at right angles a rectangular or tapered chamber is housed in the other leg of the chamber. Another combination is to use the same chamber for far-field and near-field testing.

Again, this combination permits both the measurements of small- and large-aperture antennas in a common anechoic chamber. Special care is required in designing these facilities to ensure that compatible operation is achieved over the entire operating range.

8.5.2 Design Considerations

The anechoic design considerations mainly consist of making sure that the lowest frequency of operation is properly budgeted and that the design is compatible with the other modes of operation.

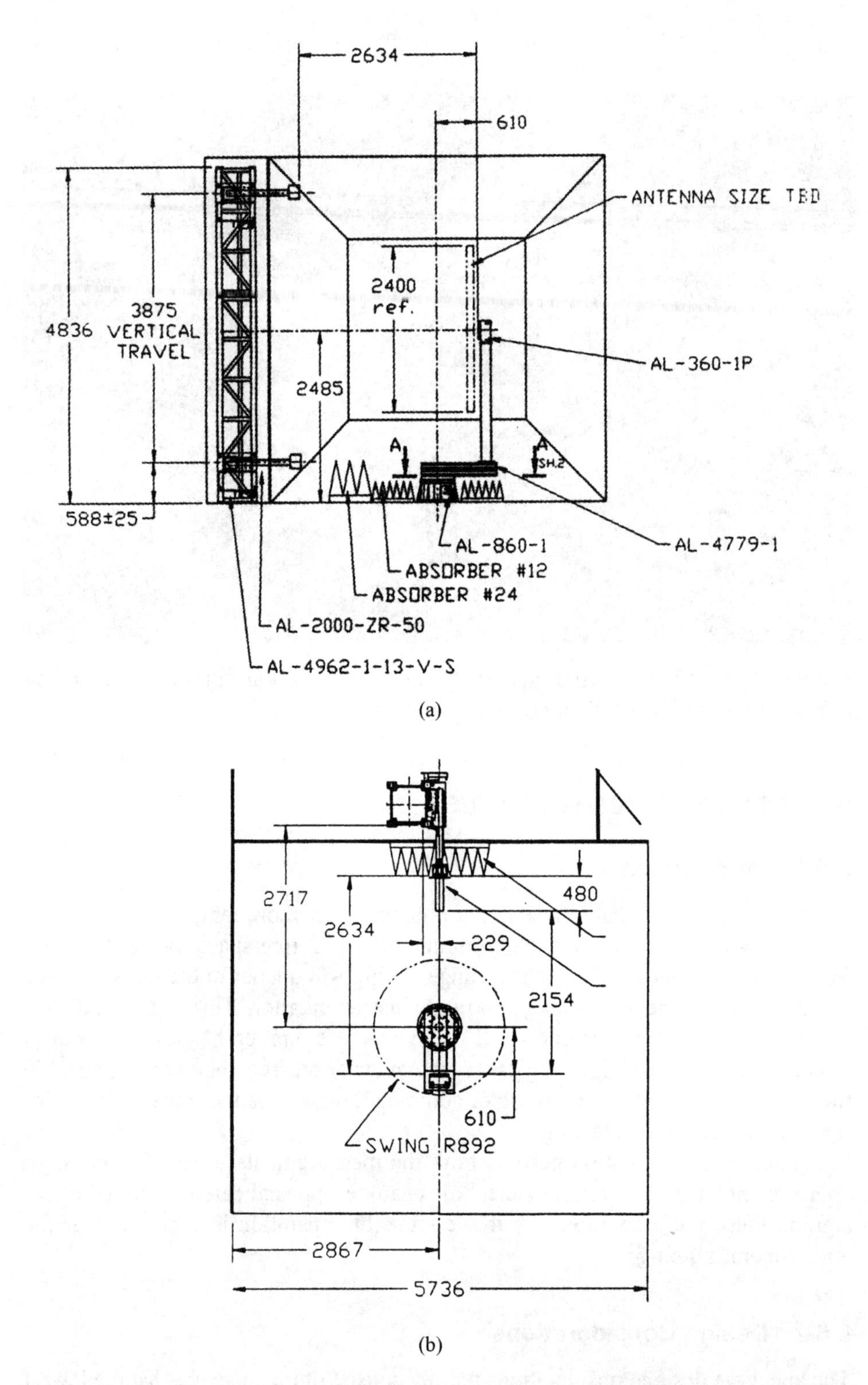

Figure 8.25 (a) Consolidated tapered chamber and (b) cylindrical near-field chamber. The dimensions are given in millimeters. The scanner is located in the side wall of the chamber's test region. (Drawings courtesy of Advanced Electro-Magnetics, Inc., Santee, CA.)

8.5.3 Design Examples

One recent example has been reported. It is a combination of a tapered chamber for operation down to 400 MHz and a cylindrical near-field range for larger-aperture wireless base station antennas [12]. This concept is illustrated in Figures 8.25(a) and 8.25(b). Other combinations are possible. Care must be taken to ensure that the various operating modes are mutually compatible.

8.5.4 Acceptance Test Procedures

The standard chamber acceptance procedures apply for each of the modes of operation.

8.6 THE TEM CELL

8.6.1 Introduction

A mini version of an anechoic chamber is the TEM cell. Crawford originally conceived the transverse electromagnetic mode cell at the National Bureau of Standards in the early 1980s [13]. The Crawford cell, or TEM cell, consists of a section of rectangular transmission line tapered at each end to a transition section that includes standard coaxial connectors. Careful design and construction can ensure accurate field measurements within ±1 dB while still maintaining 50-ohm input impedance at both ends of the cell. The cells are equipped with hinged doors that provide clearance for objects up to the size of the test region within the cell. Additional ports are provided for installation of probes and connectors. The cell offers an efficient means of accurate and broadband field measurement for testing printed circuit boards, and models are available which can be used up to desktop computer-sized equipment. Anechoic absorbers are used in the cell to reduce higher-order modes.

Applications include:

a. Susceptibility testing
b. Emissions testing
c. Meter/sensor calibration
d. Product calibration
e. High-power leveling

A variety of cell configurations have evolved. The original geometry has become popular for small items, such as circuit cards or modules. The gigahertz TEM cell (GTEM) [14] is widely used to measure the susceptibility and emission properties of a variety of electronic equipment. Another version is the EUROTEM® [15], which provides a means of switching between vertical and horizontal polarization without rotating the device under test, as required in the normal TEM configuration.

8.6.2 TEM Principles of Operation

The TEM cell was developed to meet the need for electromagnetic field testing of electronic equipment. The uniform field in a TEM cell is achieved by using a stripline-type center conductor within the cell. As the frequency increases, the wavelength decreases to a point where the separation between the center conductor and the cell wall approaches one-half wavelength. When the separation approaches one-half wavelength, higher order modes other than the TE mode appear, and the field uniformity degrades.

The GTEM overcame the mode problem by offsetting the stripline and lining the interior of the cell with absorber, as illustrated in Figure 8.26. The long taper design provides for minimum distortion to the field launched at the narrow end of the cell. The absorber is designed to effectively terminate all modes at the end of the stripline. The GTEM is a patented device [15] and is manufactured under license by several companies. Figure 8.27 is a photograph of a typical GTEM system.

The EUROTEM® is a patented compact symmetrical TEM device with four striplines that are surrounded by a fully absorber-lined ferrite enclosure. It is somewhat like a four-sided GTEM, with the test region in the center of the four striplines. The switches located at the input to the transmission lines control the polarization of the electric fields. All three axes of the item under test can be exposed to the electric field, thus ensuring full exposure to potential damaging electric fields. Also, all sources of emissions can be detected in a systematic way.

8.6.3 Typical Performance

Measurements of radiated emissions with the GTEM cell are accepted by the FCC and conform to requirements of IEC 61000-4-3 (EN61000-4-3). The GTEM cell is a smaller, lower-cost alternative to anechoic chambers and OATS facilities. It permits EMC testing within a compact tapered enclosure. It is ideal for in-house test laboratories and QA test centers, with good correlation to OATS as verified and approved by many of the international authorities. Typical characteristics of the available models include:

Frequency range: DC > 1 GHz

VSWR typical: < 1.5:1

VSWR maximum: < 1.8:1

Maximum input power, CW: Ranges from 50 to 2 kW, depending on chamber size.

Test region size: 15 × 15 × 10 cm up to 98 × 98 × 55 cm.

Cell size: From 1.4 × 0.75 × 0.5 m up to 8 × 4 × 3.2 m.

Correlation for 3-, 10-, and 30-m OATS measurements, to within ±4 dB from 30 MHz to 1000 MHz.

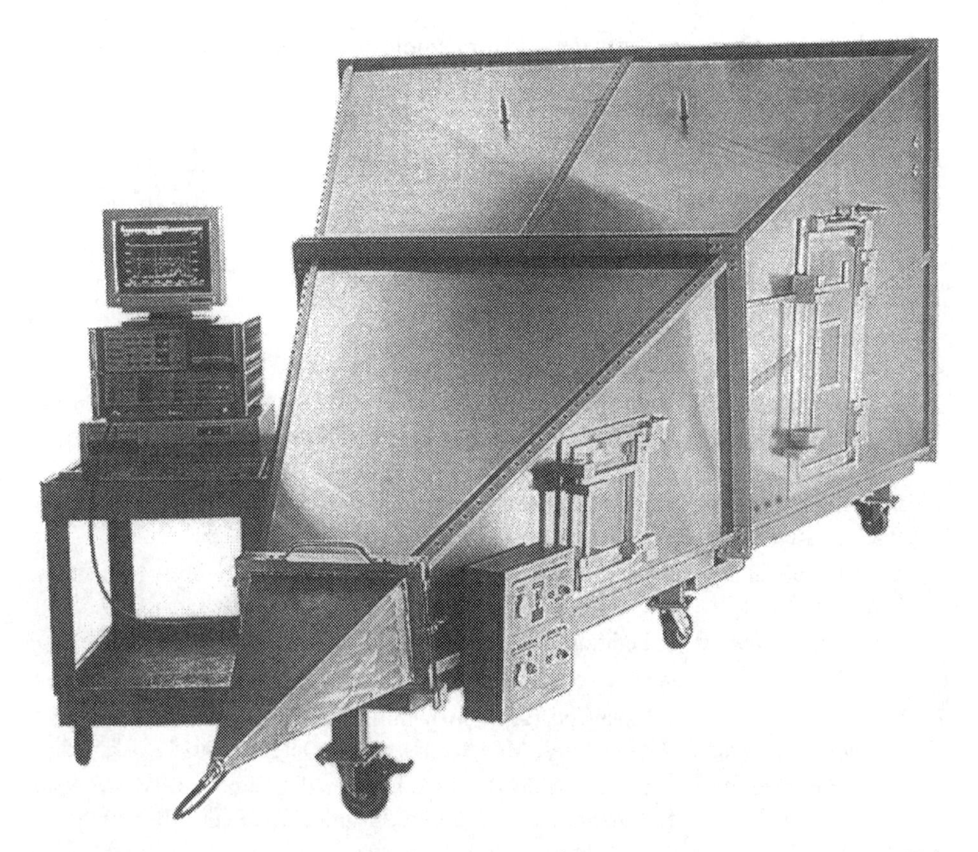

Figure 8.27 GTEM cell. [Photograph courtesy of EMC Test Systems (ETS), Austin, TX.]

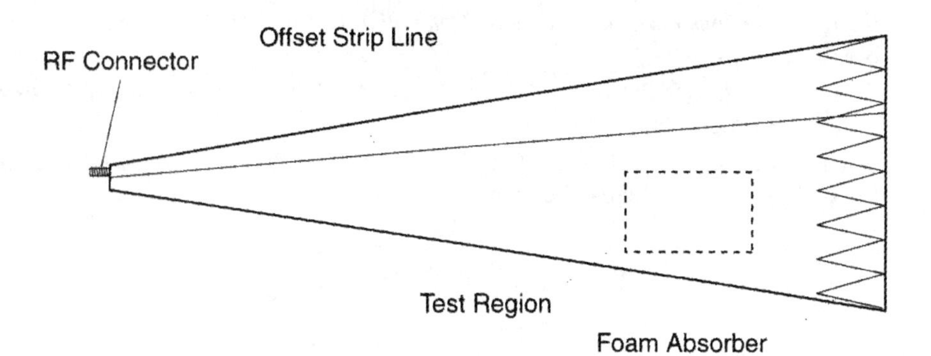

Figure 8.26 Operation of the GTEM cell.

Shielding effectiveness: Typical enclosure shielding effectiveness is as follows:

- 55 dB from 0.01 to 0.1 MHz
- 80 dB from 0.1 to 10 MHz
- 85 dB from 10 MHz to 10 GHz

REFERENCES

1. W. H. Emerson, U.S. Patent No. 3,308,463, March 1967, *Anechoic Chamber*.
2. T. G. Hickman and T. J. Lyon, *Experimental Evaluation of the MIT Lincoln Laboratory Anechoic Chamber,* Scientific-Atlanta, Inc., under AF Contract 19(628) 5167, June 1968.
3. H. E. King et al., *94 GHz Measurements of Microwave Absorbing Material,* March 1966, AF Report No. SSD-TR-66-71.
4. E. E. Jordan, *Electromagnetic Waves and Radiating Systems,* Prentice-Hall, New York, p. 611, 1950.
5. J. D. Kraus, *Antennas,* 2nd edition, McGraw-Hill, New York, p. 159, 1988.
6. A. L. Holloway, U.S. Patent No. 3,806.943, April 23, 1974, *Anechoic Chamber*
7. W. D. Burnside et al., An Enhanced Tapered Chamber Design, *Antenna Measurement Techniques Association Proceedings,* Monterey Bay, CA, October 1999.
8. W. D. Burnside et al., An Ultra-Wide Bandwidth Tapered Chamber Feed, *Antenna Measurement Techniques Association Proceedings,* Seattle, WA, September 1996.
9. J-R. J. Gau et al., Chebyshev Multilevel Absorber Design Concept, *IEEE Transactions on Antennas and Propagation,* Vol. 45, No. 8, August 1997.
10. L. H. Hemming, U.S. Patent No. 4,507,660, March 26, 1985, *Anechoic Chamber*.
11. G. A. Sanchez, U.S. Patent No. 5,631,661, May 20, 1997, *Geometrically Optimized Anechoic Chamber*.
12. V. Harding et al., A Cellular Band Far-Field and Cylindrical Near-Field Tapered Anechoic Chamber, *Antenna Measurement Techniques Association Proceedings,* pp. 37–43, October 2000.
13. M. L. Crawford, Generation of Standard EM Fields Using TEM Transmission Cells, *IEEE Transactions on Electromagnetic Compatibility,* EMC-16, No. 4, pp. 189–195; November 1974.
14. D. Hansen et al., U.S. Patent No. 4,837,581, June 6, 1989, *Device for the EMI Testing of Electronic Systems*.
15. D. Hansen et al., The EUROTEM®, a New Symmetrical TEM Device, data sheet of EURO EMC Service, Teltow, Germany.

CHAPTER 9

TESTING PROCEDURES

9.1 INTRODUCTION

A variety of electromagnetic test procedures are used in the evaluation and acceptance testing of electromagnetic anechoic chambers. The microwave chamber is generally evaluated using the Free-Space VSWR Method, whereas EMI chambers are evaluated based upon procedures established by various government agencies for electromagnetic emission and immunity standards. Each of these procedures will be reviewed, along with an assortment of other test procedures that are used in acceptance testing of electromagnetic anechoic chambers.

9.2 ABSORBER TESTING

9.2.1 Introduction

Performance testing of absorbers is necessary to ensure that the manufacturing process is under control. Slight changes in the chemistry of the material used in manufacturing absorbers can affect electrical performance of the absorbing material. It is common to conduct 100% testing of the product at a minimum of one frequency. It is recommended that this frequency be selected so that the material is on the order of one wavelength thick, as discussed in Chapter 5. Ferrite material requires the use of a waveguide or rectangular coax test fixtures because of its lower frequency of operation.

9.2.2 Testing of Microwave Absorber

Microwave absorber testing is generally conducted using an arch system. A pair of broadband antennas is located above a conductive table. The antennas are located as near normal to the plate as is physically possible. This arrangement approximates the measurement of the normal incident reflectivity of the absorber being tested, illustrated in Figure 9.1. The procedure consists of determining a reference level by calibrating against a bare metal plate. The absorber is then placed upon the plate. The difference in the readings is the reflectivity of the absorber. A study conducted for NASA by the University of Michigan [1] determined that the absorber should be a certain number of wavelengths on a side to accurately measure a given reflectivity. It is generally best to have an absorber sample of at least six wavelengths on a side to reliably determine reflectivity on the order of –50 dB. Measurement arches to determine performance at 1 GHz are quite common. It is difficult to go below this frequency, although an arch designed for 250 MHz was used for a limited time for testing a 3.66-m (12-ft) pyramidal absorber being used in a very large rectangular chamber.

9.2.3 Low-Frequency Testing

At VHF frequencies, it is recommended that a large flared waveguide be used to conduct the reflectivity testing. Sliding load techniques are used, because reflectivity on the order of –50 dB means the resulting VSWR will be very small.

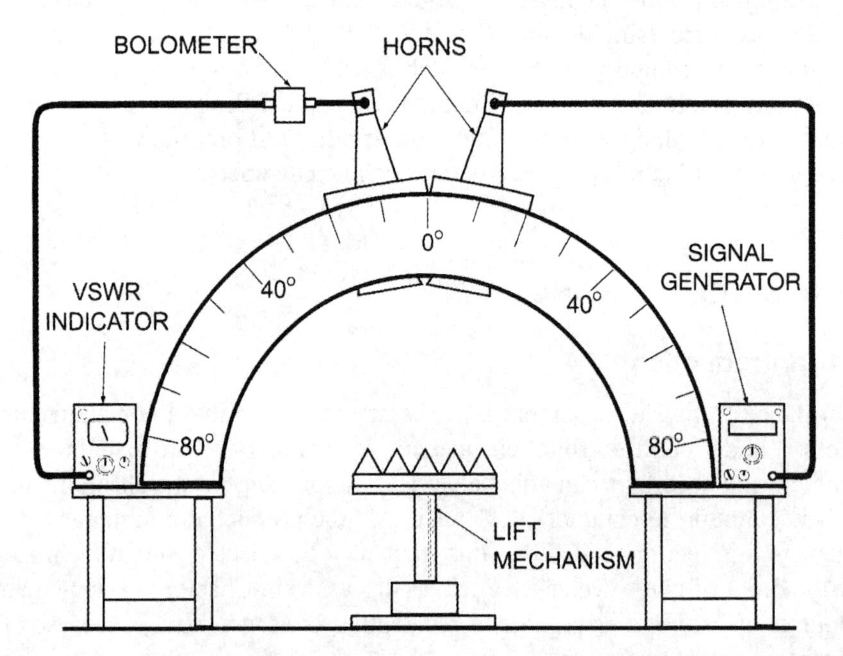

Figure 9.1 NRL arch test fixture commonly used for absorber measurements.

With the advent of the automatic network analyzer and electronic gating, it is now possible to construct reasonably sized rectangular coax test fixtures [2] that can measure down to low frequencies and very low VSWR. An excellent method to saving floor space is to place the waveguide vertically. The absorber sample can be raised into the mouth of the test fixture by using a scissors lift mechanism. Research at the National Institute of Standards and Technology (NIST) [3] demonstrated that measurements can be performed on relatively small samples of hybrid ferrite absorber using automatic network analyzers and time gating.

9.2.4 Compact Range Reflector Testing

The most common method of evaluating a compact range reflector system is to field probe the test region of the system, illustrated in Figure 9.2. Because this is the direct path field being probed, the peak-to-peak ripple is read directly off the field probe data and the extraneous signal level is read directly from the chart given in Figure 9.3.

Figure 9.2 An example of a compact range field probe fixture with absorber treatment. (Photograph courtesy of Lehman Chambers, Inc., Chambersburg, PA.)

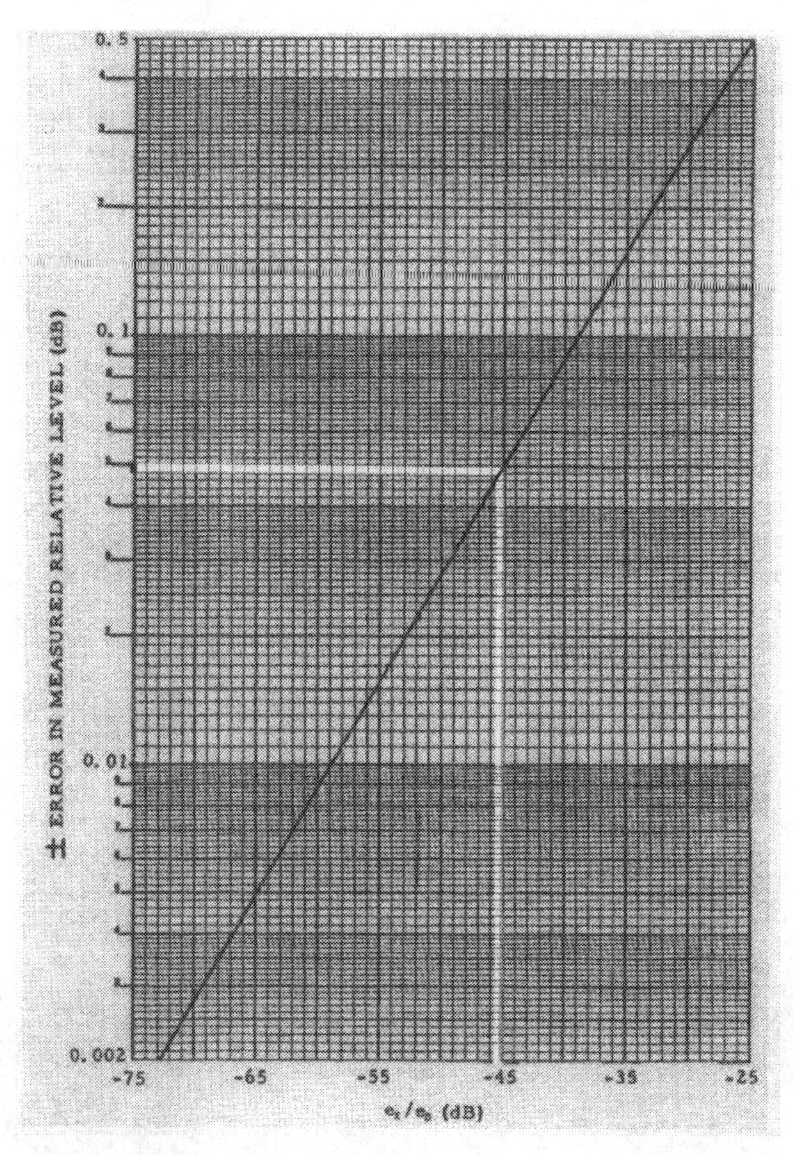

Figure 9.3 The extraneous energy level (reflectivity) versus pattern ripple and pattern.

9.2.5 Fire-Retardant Testing

Since the advent of the NRL 8093 Fire-Retardant Testing requirement [4], to which all manufacturers must adhere, only a small number of fires have occurred. Prior to the adoption of this test method, the government had several damaging fires in which personnel were injured from the toxic smoke, and large amounts of equipment were destroyed. The sources of ignition in anechoic chambers include heat from welding during chamber construction, soldering irons, high-intensity lighting, chemical reactions, arcing from power cords, high-power RF energy, sparking or heat build up in an antenna positioner, and frictional heat. Once ignited, absorber that has not been treated is highly combustible. The foam used in the fabrication of the product provides both fuel and oxygen (in the cells of the foam). Thus it becomes difficult to extinguish. It gives off toxic smoke containing carbon monoxide, hydrochloric acid, and hydrogen cyanide.

The NRL test procedure addresses most of these concerns in the five tests specified in the procedure. Tests 4 and 5 are not commonly specified, because the results have varied greatly from the various laboratories. An effort by a group at Texas Instruments, now owned by Raytheon, attempted to improve the repeatability of the tests and published their Flammability Test Procedure [5].

The following industry standards are used to establish the flammability of absorber materials:

1. NRL 8093 Test 1, 2, and 3 (Tests 4 and 5 are not used due to repeatability problems). Test 1 is resistance to electrical stress. It is conducted on three test samples measuring 5.08 cm × 15.24 cm × 15.24 cm (2 in. × 6 in. × 6 in.). Electrodes are inserted 2.54 cm (1 in.) into the absorber at a distance of 2.54 cm (1 in.) apart. The absorber is unprotected by painted surfaces. An attempt is made to initiate combustion with a 240-V, 8-A (minimum) power supply. Power is applied for 60 seconds. To pass the test, the absorber sample must fully extinguish (no visible flame, smoke, or smoldering) within 60 seconds after power is turned off. Each sample is weighed before and after the test. Not more than a 20% weight loss is permitted.

   Test 2 is ease of ignition and flame propagation. Test 2 is conducted on a minimum of three test samples consisting of 5.08-cm (2-in.) cubes, unprotected by painted surfaces. The sample must self-extinguish following a 30-seconds exposure to flame from a Bunsen burner.

   Test 3 is a modified smoldering test. Test 3 is the only recognized test to evaluate solid-phase combustion. A cartridge heater is inserted 3.81 cm (1.5 in.) into the top center of a 15.24-cm (6-in.) cube and heated to 600° C (1112° F). This temperature is maintained for 5 minutes to initiate solid-phase combustion. The heater is then removed. The material must fully self-extinguish within 30 minutes. After the test, each sample is cut open and inspected for evidence of remaining combustion.
2. Limiting oxygen index (LOI), percentage of oxygen required for combustion.
3. ASTM E84-84, flame-spread index (also known as a tunnel test) used for building materials.
4. UL 94-5VA and UL 94-5VB.
5. UL 94 HBF.
6. Texas Instrument (now Raytheon) Procedure No. 2693066.
7. MIT Lincoln Laboratory Specification MW-8-21 (Test I, II, and III).
8. DIN 410 Class B-2.

9.3 MICROWAVE ANECHOIC CHAMBER TEST PROCEDURES

9.3.1 Introduction

Several procedures are commonly employed to evaluate free-space chambers. Antenna chambers are normally checked out by using the Free-Space VSWR Method, but occasionally the Pattern Comparison Method is used. Moving a target through the test region and noting the variation in the radar return to determine the purity of the test region is commonly used to evaluate Radar Cross-Section chambers. One method uses an offset metal sphere on a foam column. Another uses a corner reflector mounted on a field probe mechanism. Recently, an image method using an X–Y scanner has been developed for evaluating microwave anechoic chambers, and details on this method will be discussed in Section 9.3.4.

9.3.2 Free-Space VSWR Method

The Free-Space VSWR Method of electromagnetic anechoic chamber evaluation has been the industry standard since the early 1970s [6]. This method uses a probe antenna drawn through the test region (quiet zone) of the anechoic chamber with a slide mechanism and appropriate test instrumentation to record the interaction of the direct path signal from the source antenna and reflections from the chamber surfaces. From the data, it is possible to determine the level of extraneous signal present in the test region and determine what impact, in terms of uncertainty, that signal will have on measurements conducted in the test region.

The original concept was derived from measurements conducted on outdoor antenna ranges. Consider Figure 9.4. This depicts two antennas located at a distance apart. The direct path signal is denoted by the term E_d, and the reflected wave from the ground is denoted as E_r. The two waves interact in the aperture of the antenna under test. Figure 9.5 shows the details of the wave interaction. An equivalent representation in vector format is shown in Figure 9.6. The three illustrations show a simplified concept of total aperture field variation caused by a single specularly reflected wave.

The relationship between the level of the signals and their impact on the uncertainty in the measured data in the test aperture is derived as follows:

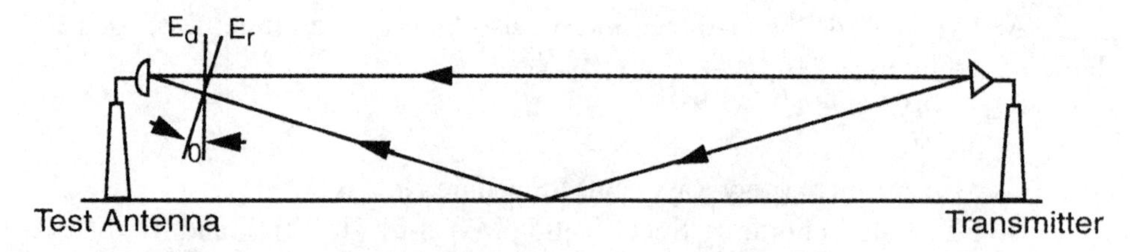

Figure 9.4 How reflected signal occur on an outdoor antenna range.

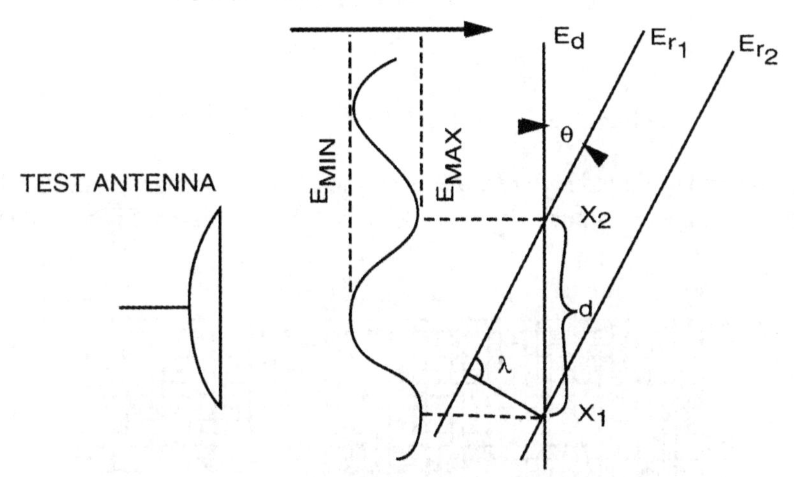

Figure 9.5 How two waves interact.

$$E_{max} = E_d + E_r \tag{9-1}$$

$$E_{min} = E_d - E_r \tag{9-2}$$

$$\Delta E = 2E_r \tag{9-3}$$

$$20 \log(E_{max}) = 20 \log(E_d + E_r) \tag{9-4}$$

$$20 \log(E_{min}) = 20 \log(E_d - E_r) \tag{9-5}$$

Let

$$\sigma = E_{max}/E_{min} = 20 \log[(E_d + E_r)/(E_d - E_r)] \tag{9-6}$$

Then

$$\log[(1 + E_r/E_d)/(1 - E_r/E_d)] = \sigma/20 \tag{9-7}$$

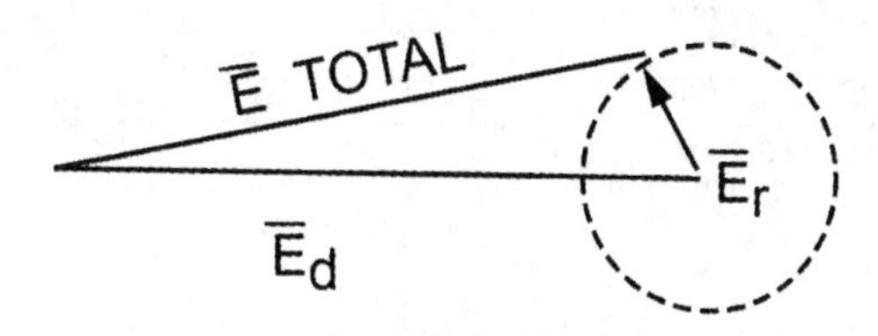

Figure 9.6 The vector representation of two interacting wave fronts.

$$[(1 + E_r/E_d)/(1 - E_r/E_d)] = 10^{\sigma/20} \tag{9-8}$$

$$E_r/E_d = (10^{\sigma/20} - 1)/(10^{\sigma/20} + 1) \tag{9-9}$$

$$20 \log(E_r/E_d) = 20 \log[(10^{\sigma/20} - 1)/(10^{\sigma/20} + 1)] \tag{9-10}$$

This equation is plotted in Figure 9.7 as possible error in measured levels relative to coherent extraneous signals, with levels from +20 to –30 dB. Figure 9.8 shows

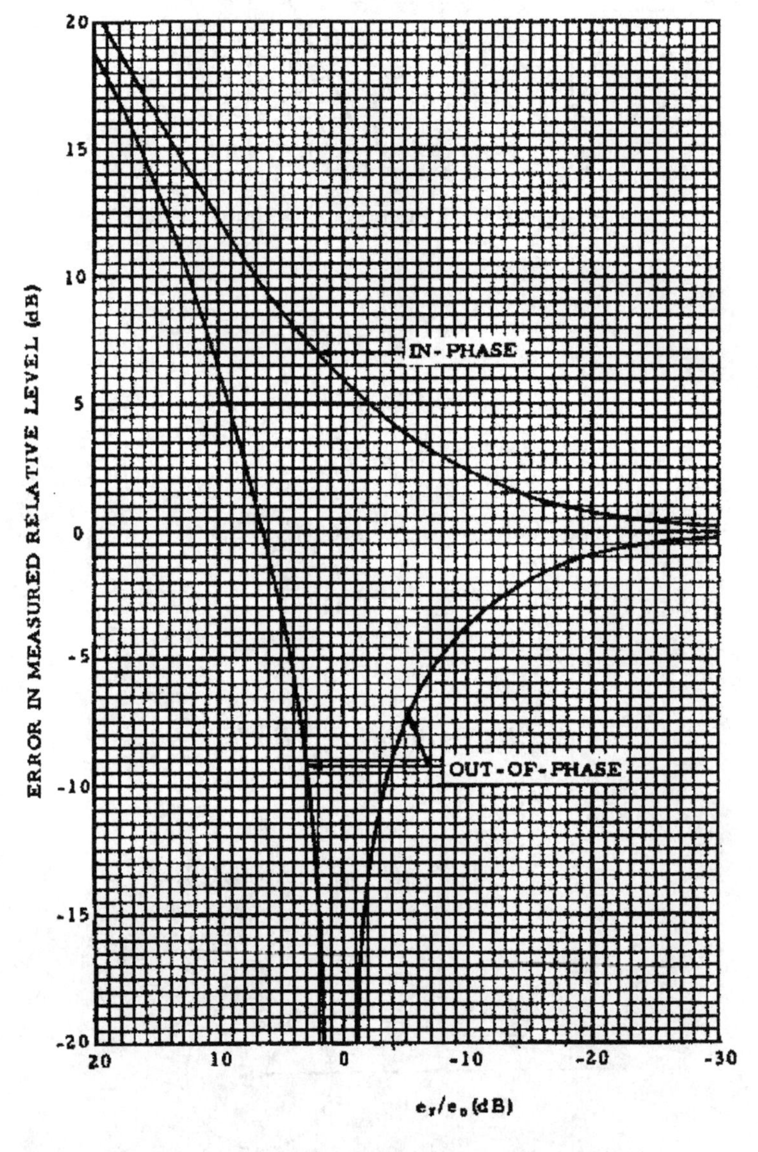

Figure 9.7 A plot of the uncertainty error versus the level of extraneous signal level from +20 to –30 dB.

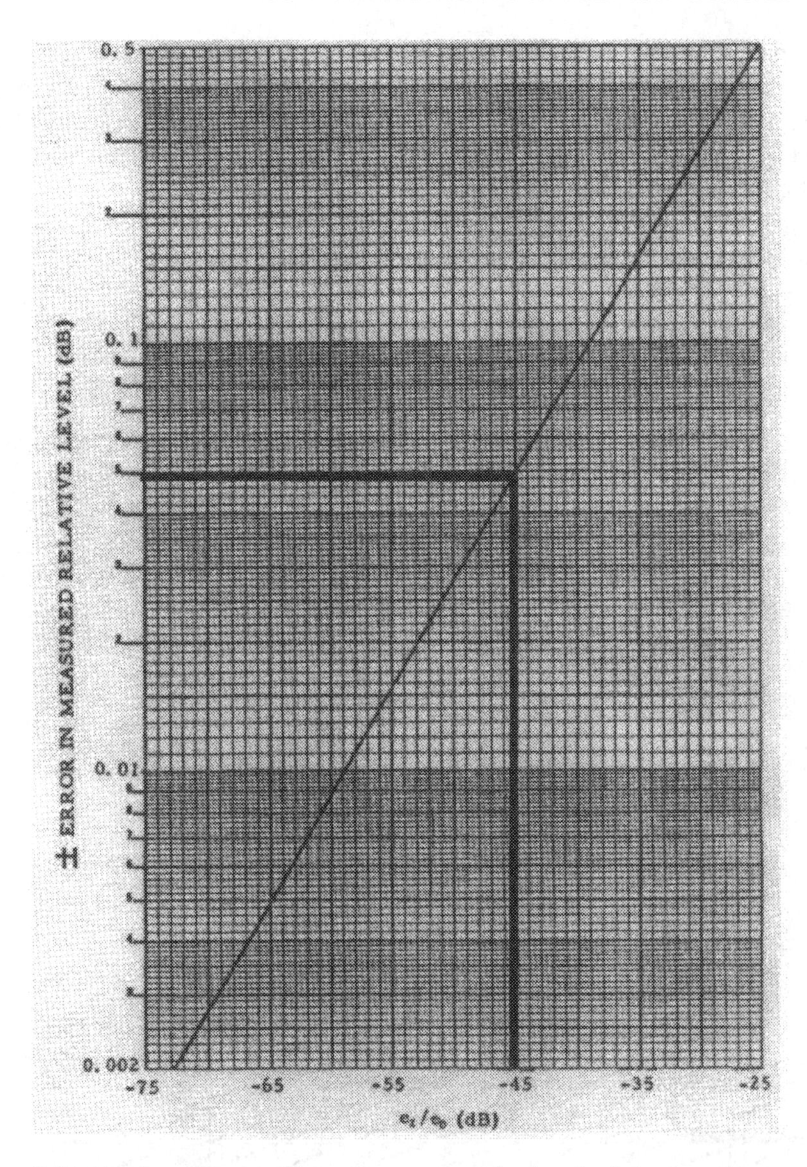

Figure 9.8 A plot of uncertainty error versus the level of extraneous signal level from –25 to –75 dB.

the same relationship for signal levels from –25 to –75 dB. The plus-or-minus errors are essentially equal for ratios of –25 dB or less, as indicated in Figure 9.8.

The above relationships assume that the signal levels arriving in the test region are received by an antenna that has an isotropic pattern (i.e., receives energy equally in all directions). When a directional antenna is used to probe the test region, the difference in the pattern level pertaining to the direct and reflected signals, respectfully, must be considered in the evaluation of the extraneous signal level. Figure 9.9 is a plot of reflectivity, as a parameter as a function of the level on the pattern of the probe versus the peak-to-peak ripple observed at that point on the pattern. As illustrated, if a 0.3-dB ripple is observed on the pattern of the probe 20 dB down from the peak of the pattern, then the extraneous signal level is equal to –55 dB. If the pattern level were –10 dB, then the measured extraneous signal level would be –45 dB.

If we know the extraneous signal level, then the uncertainty that occurs in an isotropic measurement can be determined as follows:

$$\text{Peak-to-peak ripple} = r$$
$$r = 20 \log[(1 + R)/(1 - R)] \qquad (9\text{-}11)$$

where R is the level of extraneous energy, or reflectivity, as it is commonly called in anechoic chamber design literature.

As an example, assume that $R = -45$ dB. That is, the measured level of extraneous signal was found to be –45 dB using the Free-Space VSWR Measurement Method. The amplitude of ripple this would superimpose on a direct path signal would be

$$R = -45$$
$$\text{voltage} = 0.0562$$
$$r = 20 \log[(1 + 0.0562)/(1 - 0.0562)]$$
$$r = 0.0977$$
$$\text{or}$$
$$\pm 0.0489$$

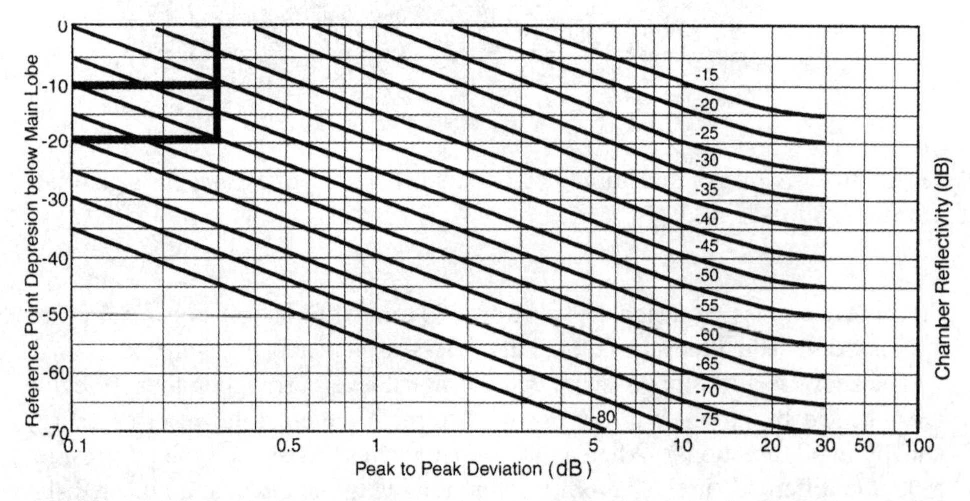

Figure 9.9 The extraneous energy level (reflectivity) versus pattern ripple and pattern level.

For convenience, the same information can be obtained using the graphs in Figures 9.7 and 9.8. If the extraneous signal level measured using the Free-Space VSWR Method found that the extraneous signal level was –45 dB, then the uncertainty on an isotropic signal would be ±0.05 or 0.1 dB peak-to-peak. This is illustrated in Figure 9.10.

In this test method a probe antenna is moved through the test region space on some type of linear probe mechanism. A directive probe antenna is the desired antenna, because when using the directional properties of the probe, an assessment can be made as to where the source and level of any extraneous energy is located. Below 1 GHz, it is common to use LPA antennas as probe antennas, while standard gain horn (SGH) antennas are used above 1 GHz.

The usual procedure is to set the probe so that a transverse scan can be made across the chamber perpendicular to the chamber axis and through the center of the test region. A reference level is established by pointing the probe horn at the source horn. Then the probe is moved across the chamber, and a recording is made of the signal variation. Two items are immediately noticed. The signal will vary as the probe moves from right to left, partly due to amplitude taper and partly due to the presence of extraneous signal levels which cause ripples in the recorded data. In high-performance chambers, the recording equipment must be able to resolve ripples less than 0.1 dB, peak-to-peak. In low-frequency chambers, the probe must move at least one wavelength through the space in order to resolve the peak-to-peak ripple caused by extraneous energy. After recording the pattern with the horn boresighted at zero degrees, the horn is rotated 10–15 degrees, and the procedure is repeated. As the horn is rotated, the pattern level on boresite will move down the pattern of the probe antenna. At any given pattern level, the peak-to-peak ripple caused by the extraneous signal is superimposed on the pattern of the probe. From the analysis given above, the level of extraneous signal occurring

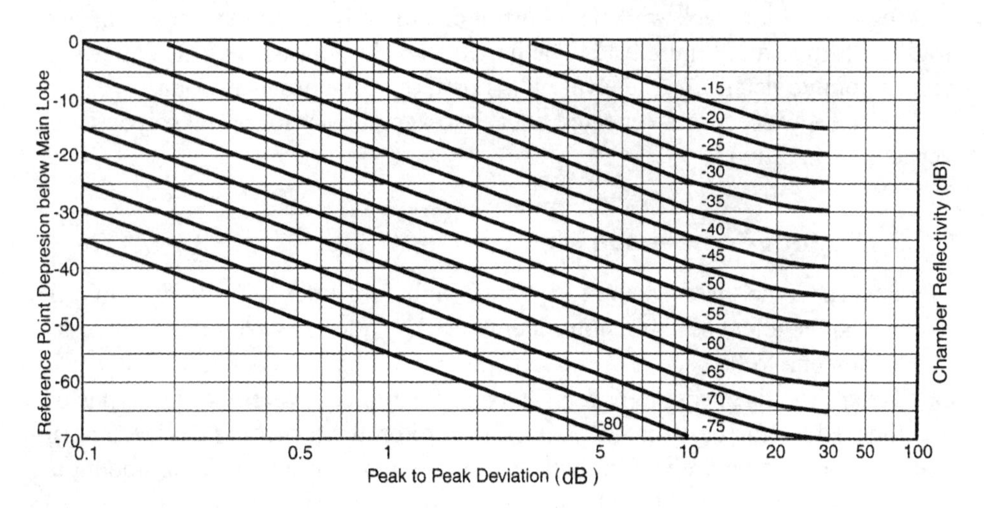

Figure 9.10 An example of the use of the reflectivity chart.

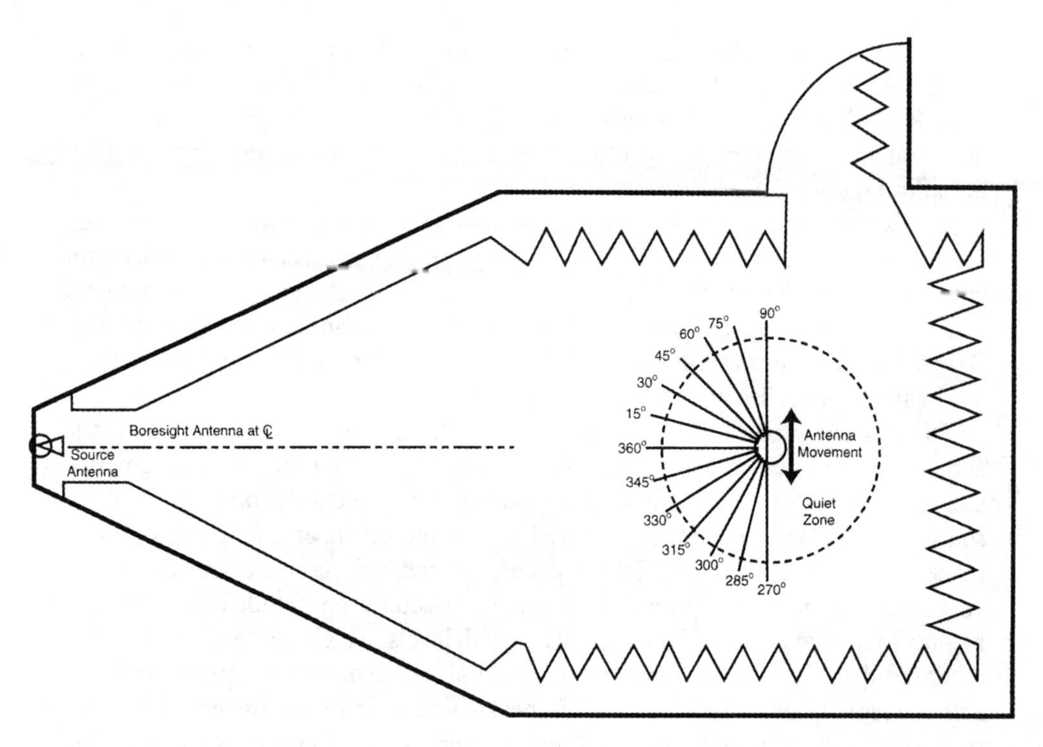

Figure 9.11 The test orientations used in the transverse test positions of the Free-Space VSWR test.

at the location the horn is pointing is determined. The rotation procedure is illustrated in Figure 9.11.

Next, the probe mechanism is set up down the axis of the chamber. The procedure is repeated by clocking the antenna from 90 degrees to 180 degrees while looking toward the back wall. The horn looks directly toward the chamber back wall as illustrated in Figure 9.12. The boresite level is still used as the pattern reference point to determine the level on the pattern of the probe antenna.

The above procedure is repeated at each test frequency for both horizontal and vertical polarization.

9.3.3 Pattern Comparison Method

The Pattern Comparison Method [6] uses a high-gain antenna (> 25 dBi) in order to evaluate the effect of extraneous energy on the pattern measurement process.

The antenna to be used as a probe is mounted such that its phase center is located over the axis of rotation of the antenna positioner. A reference pattern is cut, and the peak of the beam is normalized to zero on the recording device. The antenna and positioner are then translated along the axis of the chamber about one-eighth of a wavelength. The measurement is repeated with the peak of the beam again normalized to zero on the recording device. This procedure is repeated five

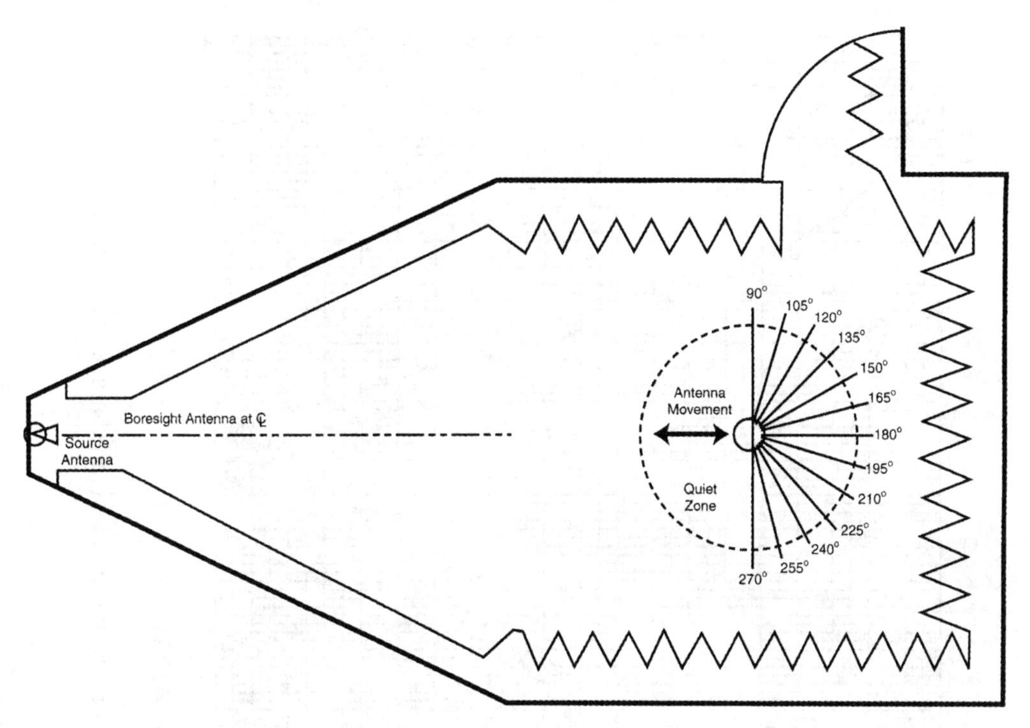

Figure 9.12 The test orientations used in the longitudinal test positions of the Free-Space VSWR test.

times while making sure that the motion translation moves through at least one-half wavelength at the test frequency. An example of this procedure is illustrated in Figure 9.13. The level of extraneous energy for the various angles of arrival in the test region are found by picking an angle on the pattern set, determining the average pattern level and peak-to-peak ripple factor, and looking up the extraneous energy level from the chart in Figure 9.9. In this case, it is possible to evaluate the effect of extraneous energy that arrives from any direction. Due to the normalizing procedure, it is difficult to evaluate the on-axis performance, from 10 to 350 degrees it is quite accurate in determining the level of extraneous energy found in the test region of a chamber when using an antenna with a half-power beamwidth less than 5 degrees. The procedure is repeated for each test frequency and for both polarizations. The result of such a test procedure is plotted as shown in Figure 9.14. This test used a moderate-gain antenna showing the difficulty to resolve the pattern changes close to the main beam.

9.3.4 X–Y Scanner Method [7]

This procedure requires use of a computer-controlled scanner. This is a portable near-field scanner with specialty software. The test system records data over the plane of the aperture under test, analyzes the data, computes the level of extrane-

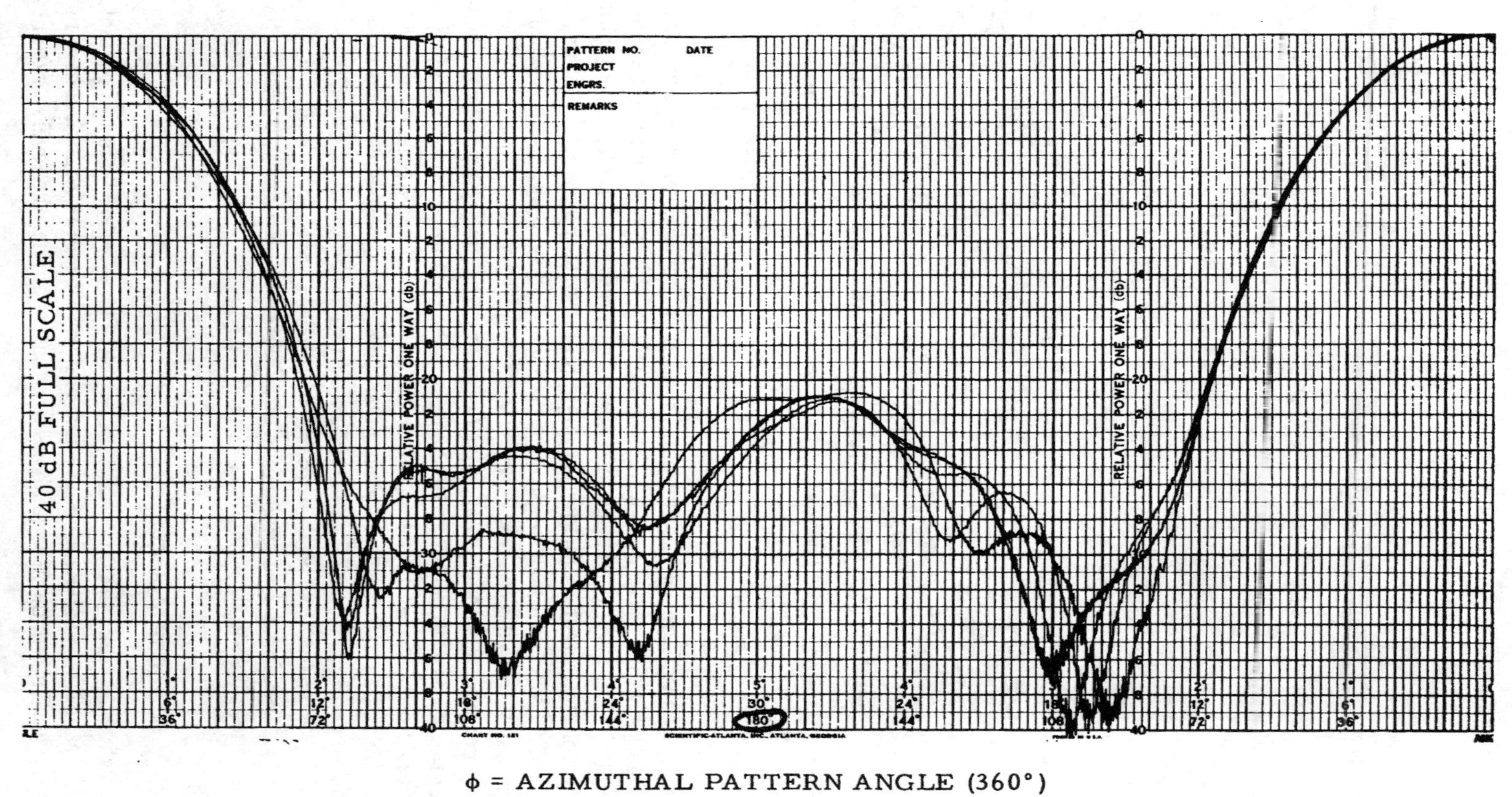

Figure 9.13 An example of the pattern comparison test procedure for anechoic chamber evaluation.

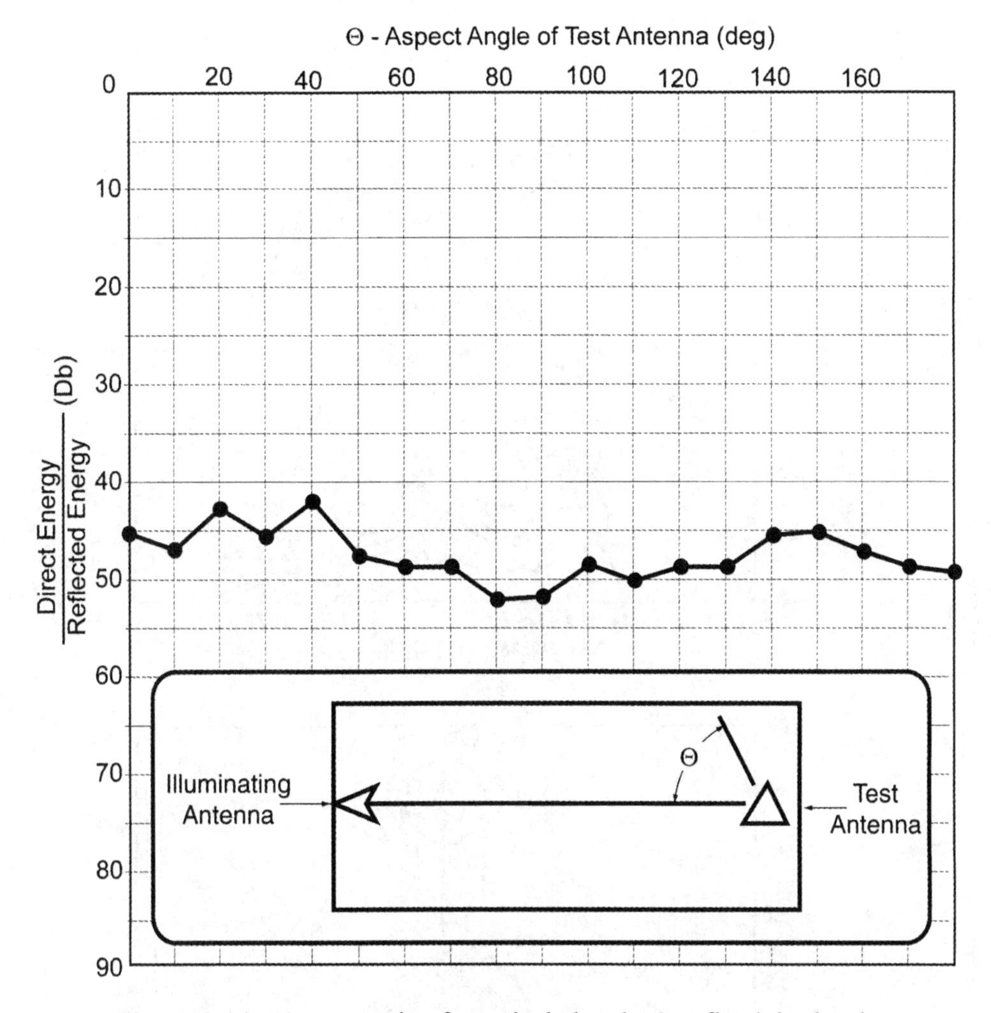

Figure 9.14 An example of a typical chamber's reflectivity levels.

ous signal level on a point to point basis, and then presents it in a graphical format, such as shown in Figure 9.15a and 9.15b.

The near-field scanner can be considered a form of two-dimensional CW syntactic aperture radar (SAR). The grid sampling points form a two-dimensional synthetic aperture antenna array. The uniformly illuminated aperture used in near-field antenna measurements must be modified into a low sidelobe design by tapering the illumination at the aperture edges. This is handled by the additional step of applying a Kaiser–Bessel or similar window function [8], depending on the spatial position of the DUT measurements.

The scan pattern can be virtually any type including planar raster, plane polar, cylindrical, and spherical [9]. The sample positions correspond to element positions in a synthetic aperture phased array antenna.

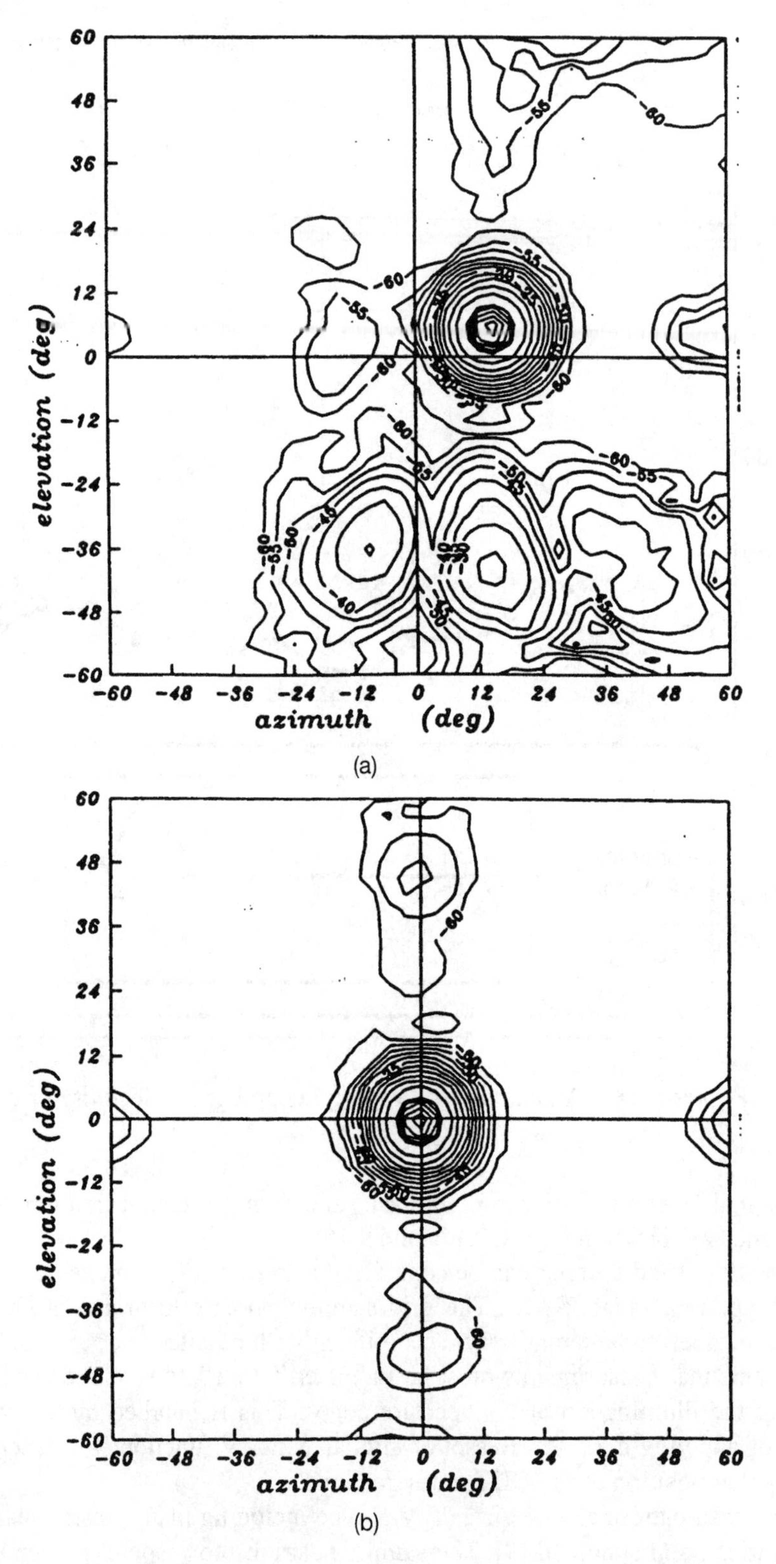

Figure 9.15 An example of the graphical data display of an X–Y scanner chamber evaluation. (a) Dirty chamber, (b) clean chamber.

Representative applications for the test system are met by measuring the phase flatness in the test region of a compact or far-field antenna range and by identifying various multipath and leakage sources.

9.3.5 RCS Chamber Evaluation

The acceptance testing of free-space RCS chambers is conducted using two different procedures. The procedure that most simulates the use of this chamber is to take a known target, such as a sphere, and move it through the test region. This is accomplished by using either a translation mechanism or a rotator, with the sphere set up on a foam column offset from the center of rotation. As the sphere is moved through the test region, the radar return from the target will vary, depending on the ripple in the illuminating field. The peak-to-peak ripple can then be related to the level of extraneous signal level in the test region. The second procedure is to use a corner reflector target that is drawn through the test region with a translation device. The pattern of the corner reflector is quite broad, and the resulting signal variation is due to the extraneous signal phasing in and out with the direct path signal. The peak-to-peak variation can be used to determine the level of the extraneous signals.

9.4 EMC CHAMBER ACCEPTANCE TEST PROCEDURES

9.4.1 Introduction

EMC chamber calibration is dictated by government standards, because the chambers are used to qualify products for emissions and immunity requirements directed by the various world governmental agencies, such as the FCC in the United States and the European Union in Europe. The two most specified procedures are the volumetric site attenuation method for emission chambers and the planar field uniformity evaluation required for immunity test chambers.

9.4.2 Volumetric Site Attenuation

Emission chambers are evaluated based on their ability to control their site attenuation to within ±4-dB uncertainty with respect to the theoretical open-area test site attenuation [10]. Briefly, the procedure is to locate a transmitting antenna in the test region and a scanning receive antenna 3, 5, or 10 m down range in the chamber. The scanning antenna is set to physically scan from 1 to 4 m in height. The instrumentation is arranged to have a max hold function so that the maximum signal is held versus frequency over the bandwidth of the antennas used in the calibration. A family of curves (Figures 9.16a and 9.16b) is generated by moving a pair of antennas with a fixed distance between them in a minimum of five locations within the test region and at two heights as illustrated in Figure 9.17. The

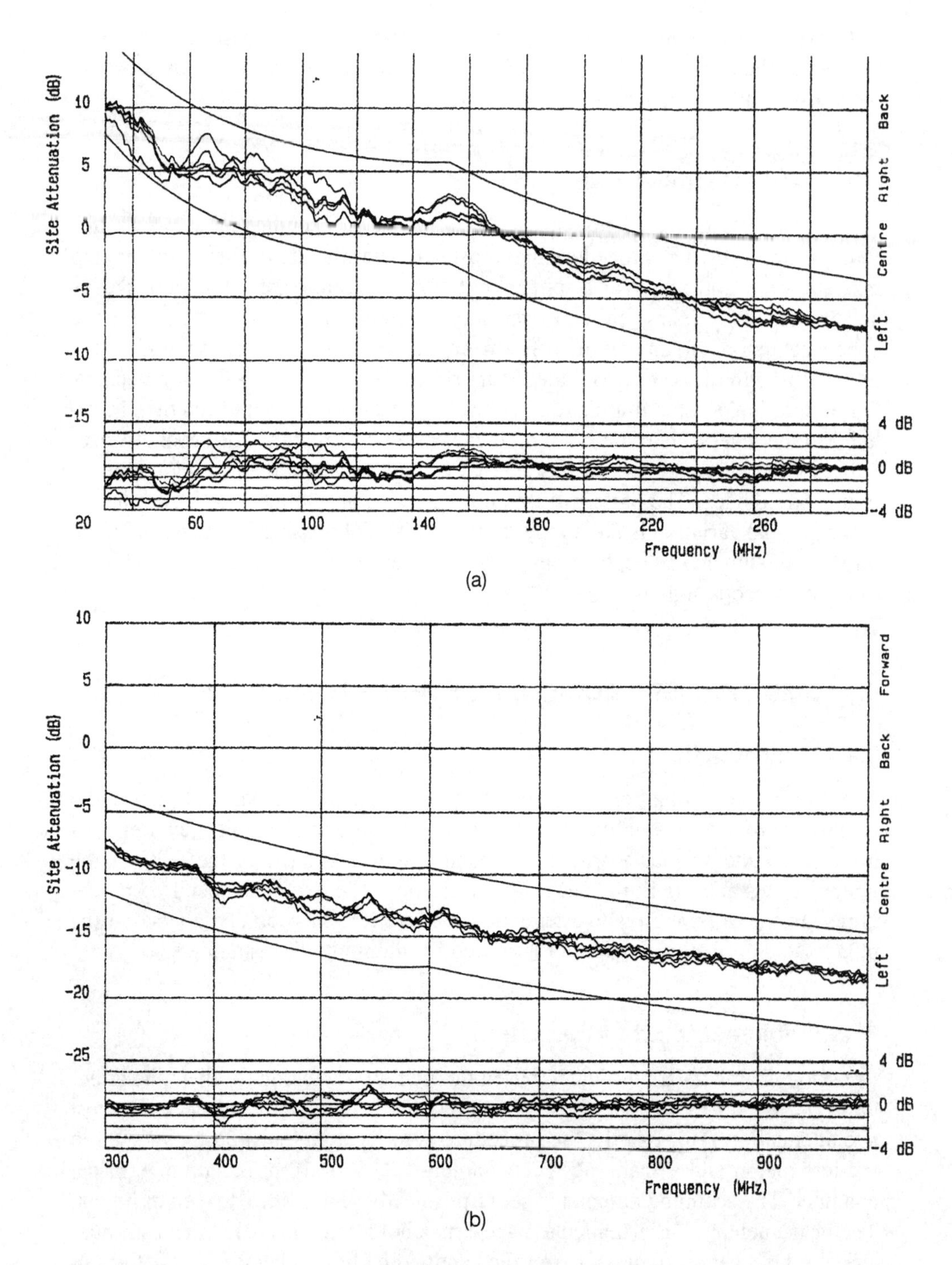

Figure 9.16 Example of measured site attenuation using the volumetric test procedure. The test band is divided into (a) a low band section using biconical antennas for the site attenuation calibration and (b) the high band portion using LPA antennas. (Example data courtesy of Panashield, Inc., Norwalk, CT.)

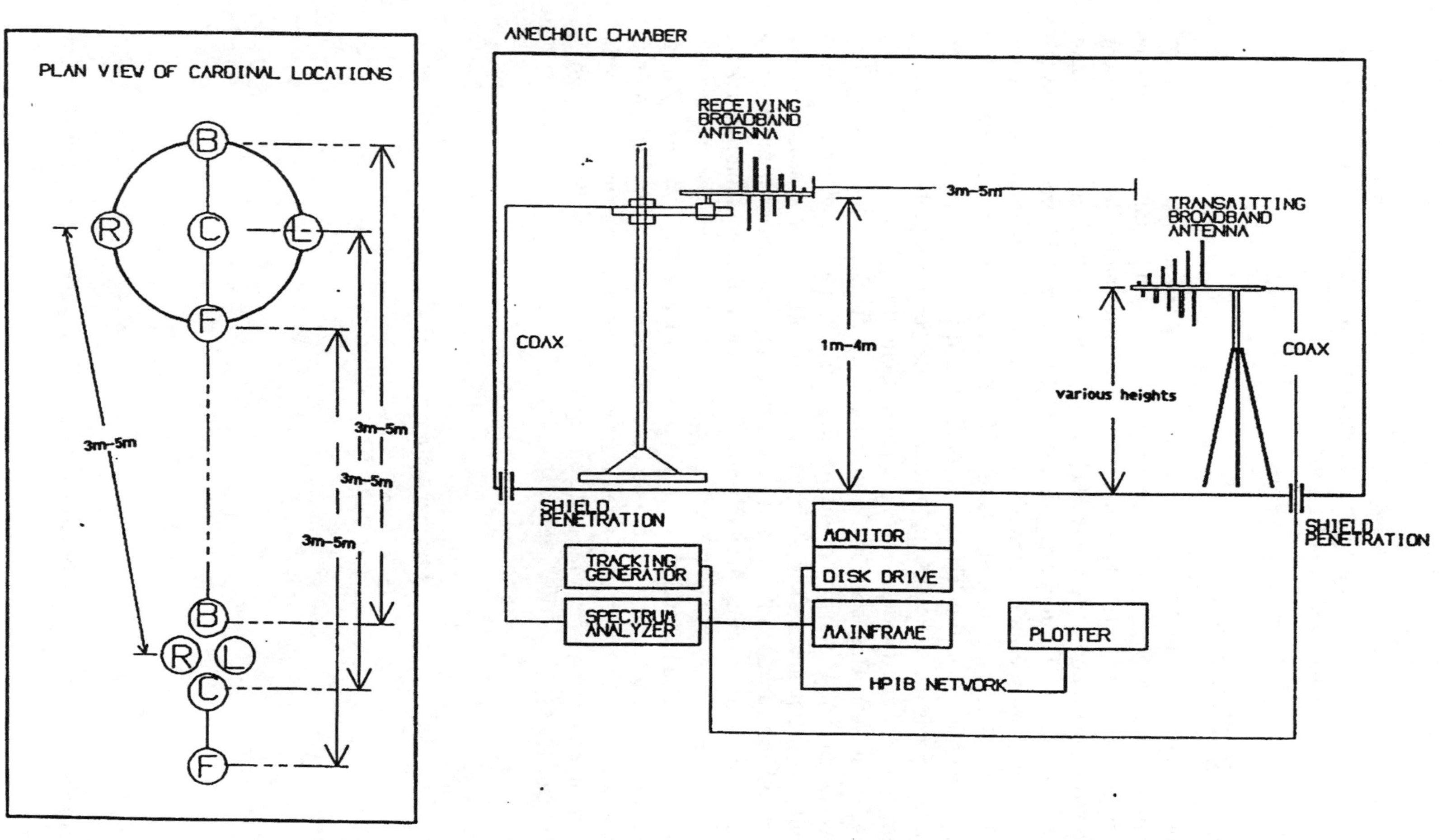

Figure 9.17 Antenna arrangement for the volumetric test procedure specified in ANSI C63.4.

separation between the antennas is held constant. The antennas should be carefully calibrated in terms of their antenna factors (AF), a parameter that includes the effect of the antenna's VSWR and gain.

9.4.3 Field Uniformity

Field uniformity is important when a device is tested for susceptibility or immunity to specific field strength. Current levels for digital equipment are on the order of 3–10 V/m up to 200 V/m for motor vehicles and military equipment. The most often quoted procedure [11] is to require the field in the test region of the chamber to be uniform within 0 to +6 dB over 75 percent of the test points within an aperture centered in the test region. The aperture is defined to be 1.5 m × 1.5 m, starting at a height of 0.8 m above the ground plane. The range length is set at 3 m. This layout is illustrated in Figure 9.18. The test procedure requires that a field strength sensor be placed at one point in the 1.5m × 1.5m grid and the field set to the required field strength over the frequency range of the test. This is accomplished using a feedback system and a leveling power amplifier. The output of the generator is computer adjusted versus frequency to maintain the field strength at a constant value. The sensor is then systematically moved to the other 15 test points, and the measurement is repeated using the same leveling function setup in the original calibration run. Twelve of the sixteen measurements shall be within the 0- to +6-dB requirement. The lowest measured level is normalized to 0 dB and the data displayed, as shown in Figure 9.19. This same procedure has been adopted for similar immunity test requirements for large-sized test regions, such as are being used in the automotive industry, where the field strength requirement has been as high as 200 V/m.

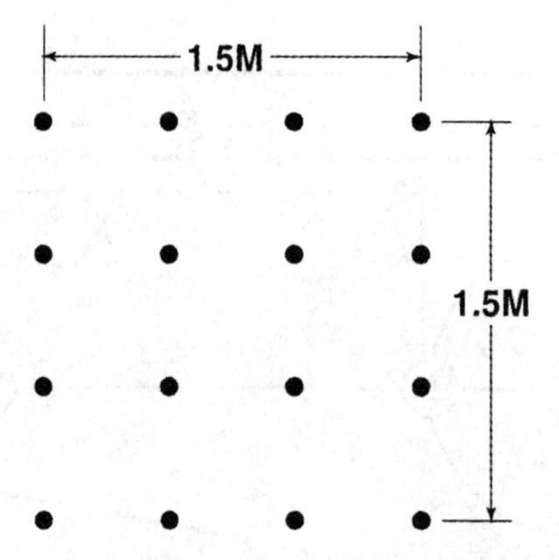

Figure 9.18 Arrangement of test points specified in the uniform field test procedure.

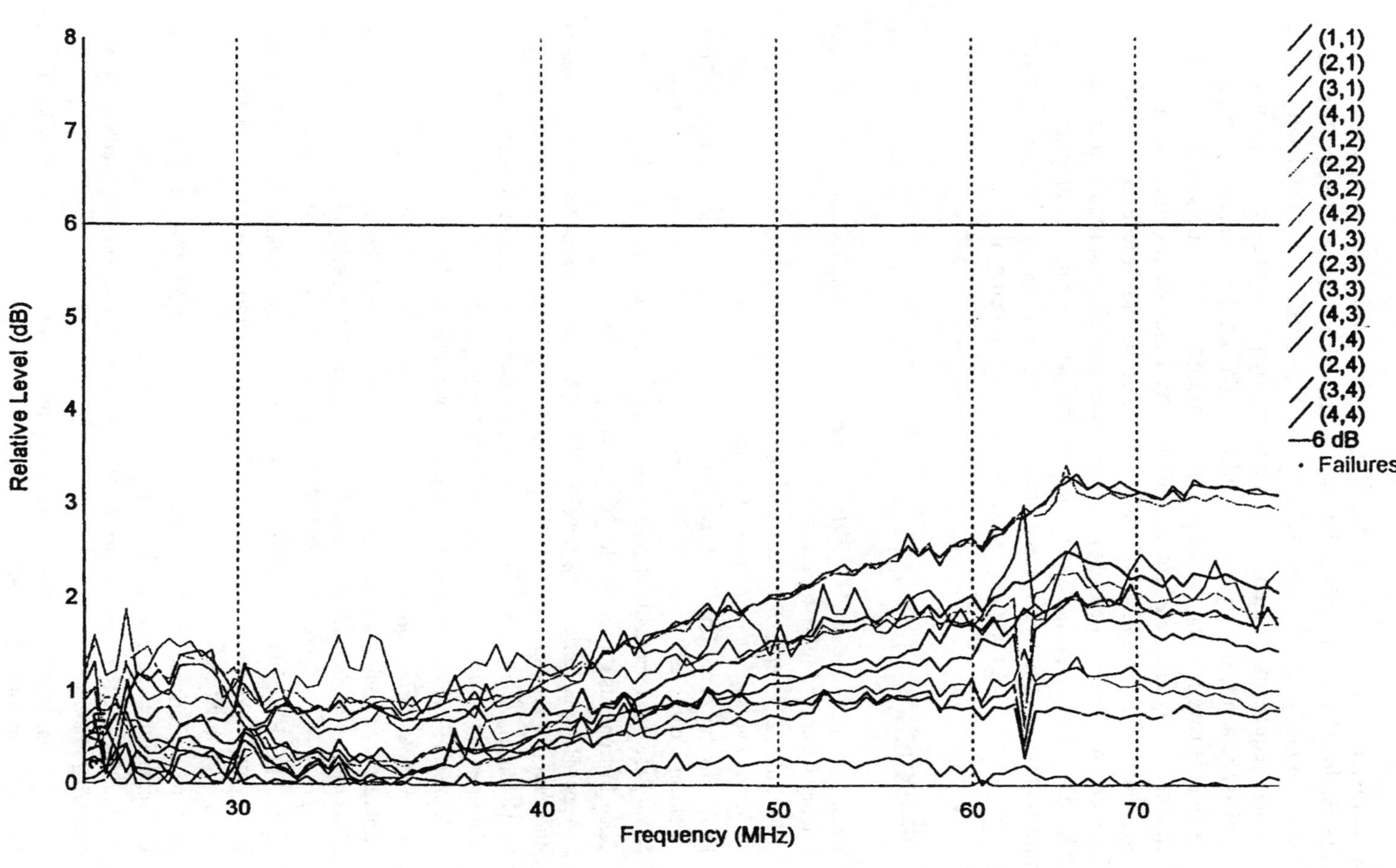

Figure 9.19 A typical plot of field uniformity test results. (Data courtesy of Panashield, Inc., Norwalk, CT.)

9.5 SHIELDING EFFECTIVENESS

The enclosures described in Chapter 4 require that once construction is completed, the shield effectiveness should be demonstrated by testing. Testing procedures that are commonly specified are NSA 65-6 [12] and IEEE 299 [13]. Each of these procedures require that the enclosure be tested by placing appropriate antennas a fixed distance apart, setting a level, and then placing the antennas on either side of the shielded wall of the enclosure. The difference in decibels is the shielding effectiveness. Magnetic, electric, and plane wave fields are measured over the frequency range of 10 kHz to 18 GHz. IEEE-299 is the most recent of the standards and is recommended for new chamber requirements. The test requirements should be tailored to clearly evaluate the shielded enclosure to ensure compliance. Care should be taken not to overspecify the requirements, since on-site testing is very expensive. A detailed review of these procedures is given in Ref. 14.

REFERENCES

1. R. E. Hiatt et. al., *A Study of VHF Absorbers and Anechoic Rooms,* University of Michigan Report 5391-1-f, February 1963.
2. IEEE Std 1128: 1998—*IEEE Recommended Practice for Radio Frequency (RF) Absorber Evaluation in the Range of 30 MHz to 5 GHz.*
3. D. R. Novotny et al., Low-Cost, Broadband Absorber Measurements, *Antenna Measurement Techniques Association Proceedings,* pp. 357–362, October 2000.
4. Naval Research Laboratory Report 8093, *Absorber Flammability Test Requirements*
5. Texas Instruments Procedure No. 2693066.
6. J. Appel-Hansen, Reflectivity Level of Radio Anechoic Chambers, *Transactions of Antennas and Propagation,* Vol. AP-21, No. 4, July 1973.
7. G. Hindman, Applications of Portable Near-field Measurement Systems, *Antenna Measurement Techniques Association Proceedings,* pp. 5-19–5-22, 1991.
8. D. Mensa, *High Resolution Radar Imaging,* Artech House, Norwood, MA, 1981.
9. D. Slater, *Near-Field Antenna Measurements,* Artech House, Norwood, MA, 1991.
10. American National Standard Institute (ANSI) C65-4-1992, *Standard for Methods of Measurement of Radio Noise Emissions form Low-Voltage Electrical and Electronic Equipment in the Range of 9 kHz to 40 GHz.*
11. CENELEC EN 61000-4-3: 1996, *Electromagnetic Compatibility (EMC)-Part 4-3: Testing and Measurement Techniques—Radiated, Radio-Frequency, Electromagnetic Field Immunity Test.*
12. NSA 65-6, *National Security Agency Specification for RF Shielded Enclosures for Communication Equipment: General Specifications.*
13. IEEE 299-1992, *Standard Method of Measuring the Effectiveness of Electromagnetic Shielding Enclosures.*
14. L. H. Hemming, *Architectural Electromagnetic Shielding Handbook,* IEEE Press, New York, pp. 143–159, 1992.

CHAPTER 10

EXAMPLES OF INDOOR ELECTROMAGNETIC TEST FACILITIES

10.1 INTRODUCTION

Electromagnetic testing takes many different forms over a broad electromagnetic spectrum. Characterizing the radiating properties of an antenna and determining the susceptibility of a computing device or to stray signals on an aircraft are but a few examples. A very large variety of test facilities have been developed to meet the needs of the electromagnetic testing community. The designs given herein are the most common of the various possible configurations used in the testing industry. Special-purpose electromagnetic test facilities are common, especially for military applications. Only a few of these are described, because the design and procurement of these types of facilities do not have a broad application. However, the principles and examples given provide general guidance for the design of just about any type of electromagnetic test facility one might encounter in practice. Pictures and performance characteristics are given for each of the more common types of indoor electromagnetic test facilities.

10.2 ANTENNA TESTING

10.2.1 Introduction

The testing of devices specially designed to radiate electromagnetic energy is extensive. This ranges from the hand-held telephones to giant antennas used in radar and astronomy. Most indoor testing, with a few exceptions, is limited to moderate-sized antenna apertures.

10.2.2 Rectangular Test Chamber

10.2.2.1 Introduction. The most common form of indoor electromagnetic test facility is the rectangular chamber. Depending on the testing requirements, these vary from a small office-sized chamber to one capable of testing a large bomber. All chambers are lined with radio-frequency absorbing material. The lining consists of ferrite tile absorber and/or hybrid versions of it for low-frequency testing. The lining consists of a base of ferrite tile and a top of foam absorber. Foam absorber alone is used in the microwave frequency range usually above 1 gH. If high-power testing is to be conducted within the chamber, then the chamber is shielded. (i.e., the enclosure is shielded to contain radiated electromagnetic energy). In some cases, testing is classified. Shielding is used to control radiated emissions so that external monitoring is prevented, or to prevent outside sources from distorting the test data. In other cases, where testing covers the common broadcast frequency bands, shielding is used to isolate the test environment from the ambient environment.

An antenna test chamber will should have, at most, a 2:1 aspect ratio. (i.e., range length/width of less than 2). This limitation is set by the wide-angle performance of the absorbing material. If the chamber is too narrow, then the angle of arrival with respect to the wall absorber is too close to grazing, and the absorber performance is severely degraded. This limitation is discussed at length in Chapter 3.

The performance of the side wall (including floor/ceiling) absorbers generally set the performance in the chamber environment. The performance of the absorber is selected so that energy reflected from the side wall is below a certain level. The amount of energy illuminating the side wall from the source antenna and the degree of extraneous energy that can be tolerated in the test region set this level, all of which are frequency-sensitive. Generally speaking, the chamber's performance is determined by the amount of extraneous energy that can be tolerated in the test region at the lowest frequency of operation. As was seen in Chapter 3, the performance of foam absorber increases with frequency, that is, the performance is a function of the thickness of the absorbers in wavelengths. When the pyramids become very large in terms of wavelength, tip scattering occurs and the normal incidence performance of the absorber declines. This generally occurs when the absorber is over 20 wavelengths in length.

10.2.2.2 Typical Rectangular Anechoic Chamber. To gain some insight into the issues involved in the design or procurement of a rectangular chamber, consider a typical sized chamber. A very common chamber is 3.66 m (12 ft) high by 3.66 m (12 ft) wide by 8.2 m (25 ft) long. The interior of the chamber is lined with a 0.3-m (1-ft) of pyramidal absorber on the transmitting wall, a 0.45-m (1.5-

ft) absorber on the receive wall, and 0.45 m (1.5 ft) of material in diagonal patches on the side walls, ceiling, and floor. The remainder of the surfaces is covered with 0.3 m (1 ft) of pyramidal material. The side-wall (ceiling/floor) design includes patches of absorber that have been rotated 45 degrees with respect to the axis of the chamber. Experimentally, it has been found that the wide-angle performance of the absorber improves about 3 dB when the grazing incident energy sees the edges of the pyramidal material and thus reduces the forward-scattered energy into the test region.

These chambers are used for routine testing and development of low-gain antennas such as dipoles, spirals, helixes, horns, and log periodic antennas. They are especially useful for measuring horn antennas used in electromagnetic countermeasure systems. The source antenna is commonly a standard gain horn antenna. The test region is 1.22 m (4 ft) in diameter and located 5.49 m (18 ft) from the transmit wall.

The test region or quiet zone may be designed to be 1.22 m (4 ft) in diameter in terms of the level of extraneous signal level expected in the zone, but the actual test diameter is set by the range equation $R = 2D^2/\lambda$. The rectangular chamber is rarely used below 0.5 GHz for antenna measurements because the absorber thickness sets a practical limit in terms of size and cost. A large rectangular chamber [5.49 m (18 ft) × 5.49 m (18 ft) × 9.15 m (30 ft)] is illustrated in Color Plate 1 (refer to the special color section). Note the installation of the thick absorber in the side wall (specular region) of the chamber.

10.2.3 Tapered Anechoic Test Chamber

The tapered anechoic test chamber evolved from the ground reflection range. It was recognized that for lower-frequency broad-beamed antennas, adequate suppression of the wall reflections in rectangular chambers was not possible. Thus, a chamber design using the side-wall energy to control the illumination of the test region was created by using a conical or funnel shape around the source antenna. A large example of this design [7.32 m (24 ft) × 7.32 m (24 ft) × 18.28 m (60-ft)] is illustrated in Color Plates 2 and 3. A more typical chamber cross section is 3.66 m (12 ft) by 3.66 m (12 ft), the rectangular test region is 4.88 m (16 ft) long, and the tapered end is 5.03 m (16.5 ft) long. To make sure that the axial ratio of the field in the test region is within acceptable limits, the antenna is housed in a 1.83 m (6 ft) conic section. A square-to-round transition section about 1.83 m (6 ft) long completes the design of this chamber. The back wall is covered with 1.22 m (4 ft) pyramidal material, and the side-wall patches are 0.46 m (18 in.) thick. Wedge material is used elsewhere in the side wall including the taper section. A variety of materials are used in the cone section. The choice is usually determined by the suppliers design experience. The test region is on the order of one-third of the cross section, or 1.22 m (4 ft).

Another example of the tapered chamber is illustrated in Figure 10.1. This chamber is constructed of prefabricated shielding panels, including the tapered

Figure 10.1 Typical large shielded tapered chamber. [Photograph courtesy of EMC Test Systems (ETS) Austin, TX.]

section. Figure 10.2 illustrates the use of a 3.66-m (12 ft) twisted pyramidal absorber in the back wall of a tapered chamber.

10.2.4 Compact Range Test Chamber

The compact range concept utilizes especially designed reflectors to achieve a uniform field while illuminating the device under test. The most common method is to use a specially designed large reflector to establish a uniform field in amplitude and phase, in the radiating near field of the reflector. This technique permits the testing of large electrical apertures within an indoor test facility. As an example, testing a 1.22-m (4-ft) aperture operating at 18 GHz requires a 182.9-m (600-ft) outdoor range. A compact range reflector on the order of 2.44 m (8 ft) high by 3.66 m (12 ft) wide in a room 4.88 m (16 ft) high by 6.1 m (20 ft) wide by 12.2 m (40 ft) long can perform the same test to comparable accuracy. A larger installation [13.72 m (45 ft) wide, 9.15 m (30-ft) high, and 21.34 m (70 ft) long) is illustrated in Color Plate 4.

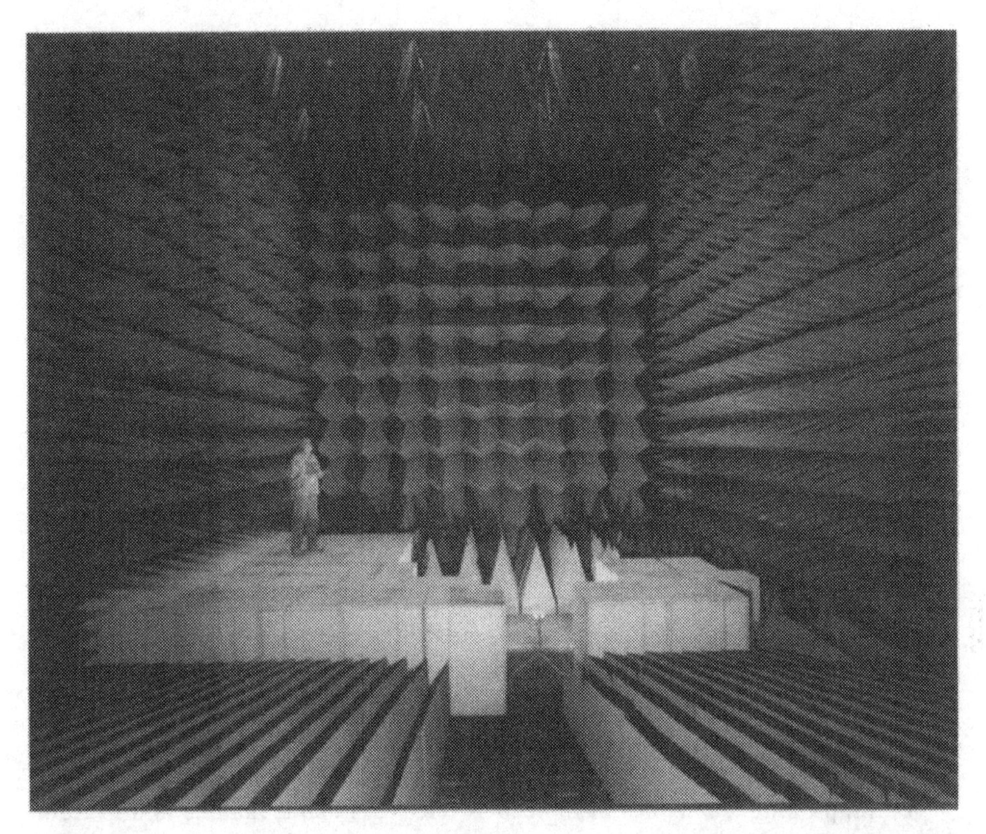

Figure 10.2 Back wall of large tapered chamber using 3.66 m (12-ft) of twisted pyramidal absorber. (Photograph courtesy of Advanced ElectroMagnetics, Inc., Santee, CA.)

A variety of reflector systems have been developed to achieve the necessary conditions for compact range operation [1]. The most common is the offset prime focus reflector system. The edges of the reflector are either rolled or serrated (see Color Plate 4) to break up the edge diffraction that would otherwise be diffracted into the test region and contaminate the measurements. The phase depends on converting the spherical wave front from the prime focus feed to a uniform phase by means of the parabolic reflector. The amplitude properties are set by the beamwidth of the feed placed at the focal point of the reflector.

A second form is the dual parabolic cylinder system. The two reflectors are arranged so that the phase front in the test region of the second reflector is planar. The amplitude properties are again controlled by the feed system. An example of this arrangement is shown in Color Plate 5.

A third form is the Cassegrain-type antenna system. The feed and subreflector are used to control the illumination of the main reflector so that the edge illumination is significantly reduced with respect to the main body of the reflector. Shaping the subreflector and the main reflector so that a planar wave front is achieved in the test region controls the phase function in the test region. The type of feed used to illuminate the sub-reflector controls the amplitude characteristics. Color Plate 6 illustrates this form of compact range. The performance of this range is similar to that specified in Table 10.2.

The energy off the main reflector is highly collimated in all forms of the compact range. The anechoic chamber design primarily involves the selection of the proper types of absorber to achieve minimum scattered energy into the test region, as well as properly terminating the range in a receive wall design, and provides good performance down to the lowest frequency limit of the reflector system.

10.2.5 Near-Field Test Chamber

The near-field method of characterizing an antenna is based upon sampling the near field of the aperture under test and then using a near-field to far-field transform to calculate the far-field pattern of the antenna. A fast receiver and a computer with a large memory are used to control the scanner collecting the data. The computer then performs the necessary mathematical operations needed to convert the data required to determine the far field. The advantage of this procedure is that the equipment can be housed within a relatively small anechoic chamber, and the technique can sample many frequencies in a single scan of the aperture under test.

Several probing techniques have been developed. Each technique is useful for different types of antennas. The most common type is the planar scanner, which samples the field of an aperture in an X–Y format, as illustrated in Figure 10.3. This method is especially useful for very large aperture directional antennas. Spherical and cylindrical scanning methods have also been developed for testing nondirectional antennas. The anechoic properties of the test chambers are selected depending on the frequency of operation of the antennas under test.

10.3 RADAR CROSS-SECTION TESTING

10.3.1 Introduction

Four facilities have been chosen to represent the state-of-the-art in radar cross-section measurements in anechoic chambers. The four are compact range test facilities. The first two compact ranges are typical of commercially available products from several manufacturers. The third is a special-purpose facility that is capable of conducting bistatic radar cross-section measurements. The geometry of the range is somewhat unique in concept. The fourth is an excellent example of the shaped reflector compact range concept, which is no longer available commercially.

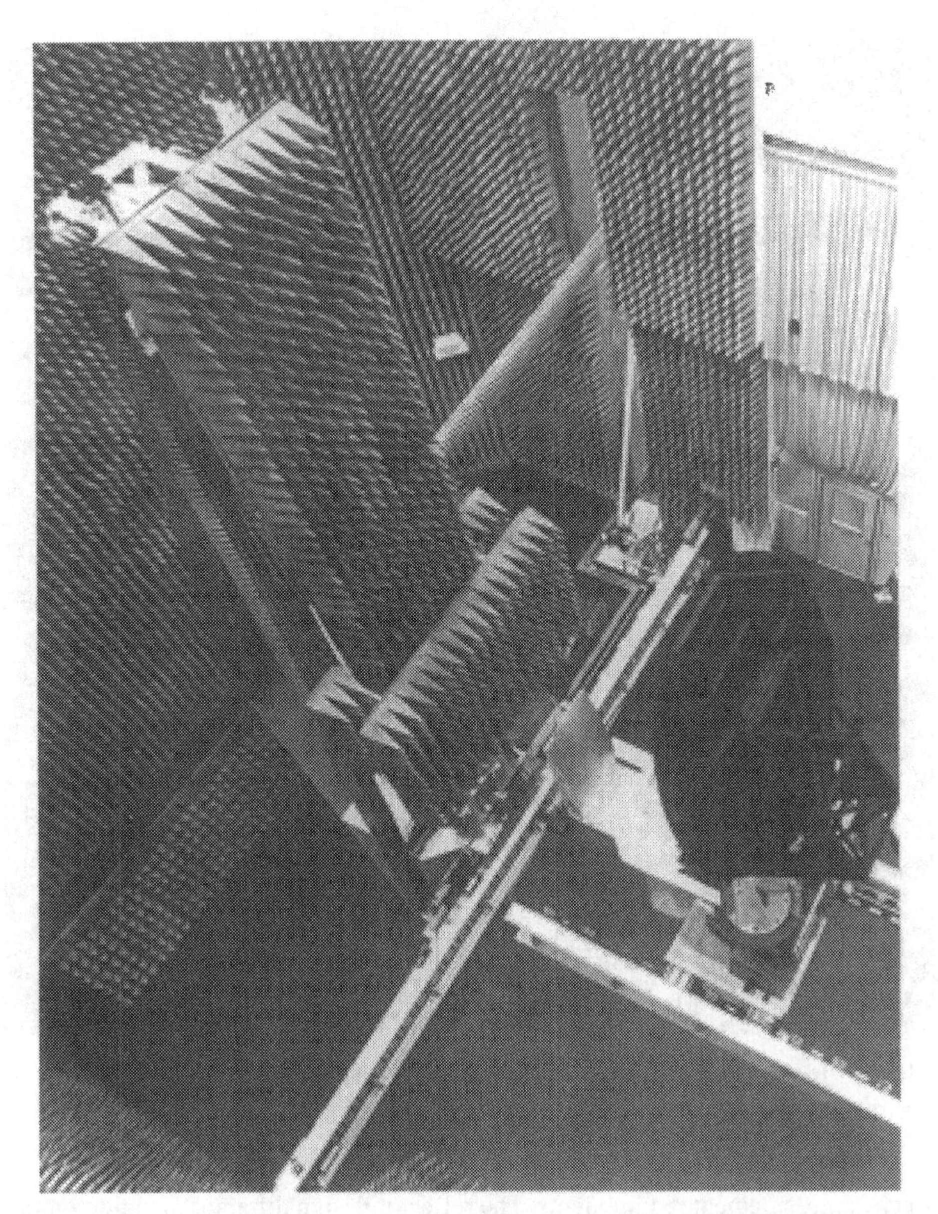

Figure 10.3 Planer X–Y scanner near-field range. (Photograph courtesy of NASA.)

10.3.2 Compact Range Radar Cross-Section Facilities

Currently, two different approaches are available commercially for implementing compact range testing. One is the prime focus approach using a single large reflector shown in Color Plate 7. This is the most common compact range geometry. Several hundred of these ranges exist throughout the world and are used for a va-

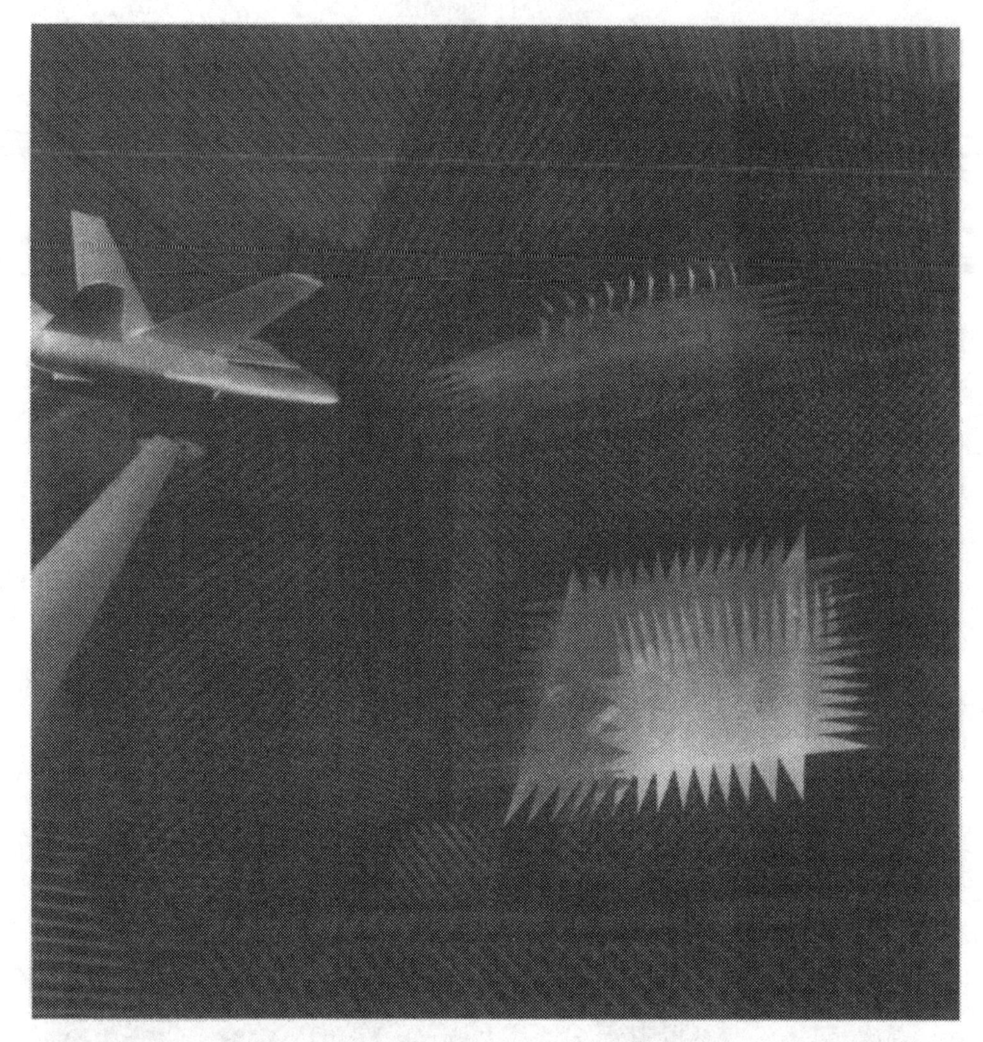

Figure 10.4 Dual reflector compact range. (Photograph courtesy of the Raytheon Company, El Segundo, CA.)

riety of measurement requirements. The reflector design differs, depending on the manufacturer. Some use serrations to terminate the reflector edges, whereas others use rolled edges [3]. Another compact range geometry consists of a dual reflector system, shown in Figure 10.4. The purpose of the reflectors is to simulate a uniform far-field condition in a short-range length. The anechoic chamber is typically rectangular and is discussed in Chapter 7.

The range capable of performing bistatic radar cross-section measurements is located at the Radar Reflectivity Laboratory at the Naval Air Warfare Center, Pt. Mugu, CA. It is housed in a large rectangular anechoic chamber. The range is illustrated in Figure 10.5. The properties and capabilities of the test facility are given in Table 10.1. The enclosure is lined with 0.91-m (36-in.)-thick pyramidal and

Figure 10.5 Bistatic compact radar cross-section range. (Photograph courtesy U.S. Navy, Pt. Mugu, CA.)

Table 10.1 Performance Properties of the Naval Bistatic Anechoic Chamber

Parameter	Data
Chamber size (L × W × H):	45.7 m (150 ft) × 45.7 m (150 ft) × 18.3 m (60 ft)
Range length:	Infinite (effective)
Range type:	Compact range
Instrumentation radar:	Lintek 5000, linear FM
Frequency coverage:	VHF, UHF, L, S, C, X, Ku, Ka, V, and W
Polarization coverage:	VV, HH, VH, HV, and circular
Dynamic range (without averaging):	> 90 dB
Range resolution (Δ_r) (depends on the bandwidth $\Delta_r = c/2B$):	< 1 cm
Angular coverage (angular separation of transmitter and receiver):	Monostatic; Bistatic: 0–90° Vertical 0–180 Horizontal
Test region (quiet zone) (L × W × H):	9.15 × 9.15 × 6.1 m (30 × 30 × 20 ft)

Source: Courtesy of the U.S. Navy, Pt. Mugu, CA.

wedge absorbers. The back wall is covered with a 1.82-m (72-in.)-thick pyramidal absorber. A second compact range reflector, not shown, is used with a separate transmitter to perform the bistatic measurements over a 180-degree arc in the horizontal plane.

Figure 10.6 and 10.7 illustrates one of the largest compact ranges in existence. This 12.2 m (40-ft) by 22.9 m (75-ft) reflector is housed in an anechoic chamber 29.3 m (96 ft) wide, 29.3 m (96 ft) high and 56.4 m (185 ft) long. The range is based upon the shaped reflector concept and has a feed turntable that automatically positions the correct feed antenna for each band of operation. This range located in San Diego, CA is used to perform large-array antenna measurements, satellite antenna measurements, and radar cross-section measurements on targets as

Figure 10.6 Large shaped reflector compact range. (Photograph courtesy of the Boeing Company.)

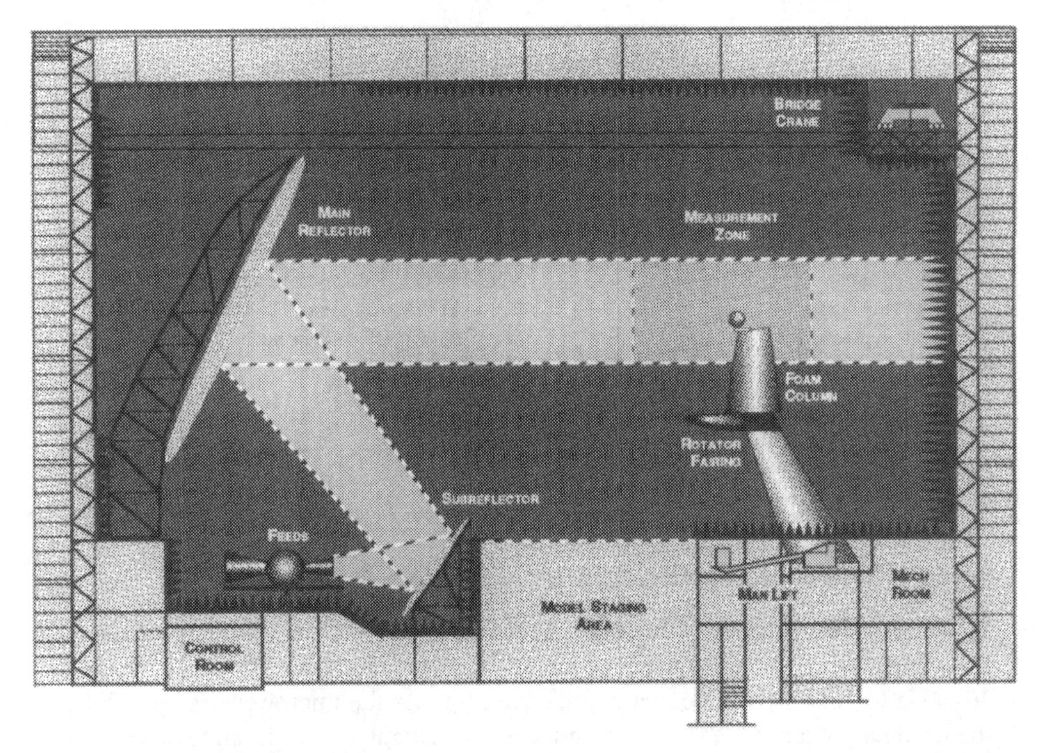

Figure 10.7 Cutaway view of large compact range. (Drawing courtesy of the Boeing Company.)

large as 8.5 m (28 ft) high by 12.2 m (40 ft) long. The chamber has 1.83 m (6 ft) twisted pyramids on the terminating receive wall, and the remaining surfaces use 0.61-m (2-ft) wedges and pyramidal material on the other surfaces. The facility test capabilities are given in Table 10.2.

10.4 EMC TEST CHAMBERS

10.4.1 Introduction

With the widespread use of electronic devices in the home, office, government and factory, it has become necessary for various governmental agencies to set up emission and immunity standards for all types of electronic equipment. This has led to the development of a series of anechoic chambers designed especially for these types of measurements. The three most common are chambers designed to perform (a) emission measurements at 3- and 10-m range lengths and (b) immunity measurements at a 3-m range length. The chamber size is dictated by the range length and by the size of the devices to be tested and by the height of the scanning receive antenna (4 m). The test items range from hand-held telephones to tractor-trailer trucks and buses. Examples of all types of EMC test facilities will be illustrated.

Table 10.2 Testing Capabilities of the Radar Measurement Facility, San Diego, CA

Parameter	Data
Chamber Size (L × W × H):	56.4 × 29.3 × 29.3 m (185 × 96 × 96 ft)
Target size (L × H):	12.2 × 8.5 m (40 × 28 ft)
Range length:	Infinite (effective)
Range type (Cassegrain):	Dual shaped compact range reflector
Instrumentation:	Hardware gated transmit/receive system
Frequency coverage:	0.3 to 18 GHz (optional 18–40 GHz)
Polarization coverage:	VV, HH, VH, HV, and circular
Dynamic range:	> 90 dB
Range resolution:	< 1 cm
Angular coverage:	Monostatic

Source: Courtesy of the Boeing Company.

10.4.2 Emission Test Chambers

10.4.2.1 Rectangular Chambers. As the speed of the computer processors reach gigahertz speeds, the need for emission testing has moved from the original 30-MHz to 1000-MHz frequency range well up into the microwave range. Thus, the need has arisen for very broadband anechoic chamber performance to support the required test requirements. As described in Chapter 5, this need is meet with various types of absorbing materials, the most common being the ferrite hybrid material. In emission testing, it is common practice to determine the level of the emission by measuring the signal level emitted by the device under test. Knowing the antenna properties and the site attenuation, the unknown level can be compared to the emission requirements. The site attenuation is a direct function of the range length for a given frequency. Thus, the signal levels on a 3-m range will be higher than those measured on a 10-m range. The extraneous signal level on the 10-m range must be suppressed more than that on the 3-m range to maintain the same level of measurement uncertainty, generally about 6 dB more. A large variety of materials are available for 3-m test sites, but only a few reliable materials are available for use on 10-m test sites. Both semianechoic (five-sided) and fully anechoic chambers are used for EMC emission testing.

A variety of test chambers are available from the anechoic chamber industry. A selection of the commercially available chambers is illustrated as a function of range length.

Three-Meter Chambers. An excellent example of the 3-m facility is shown in Figure 10.8. This chamber meets the ANSI C63.4-1992 using antennas that were calibrated using the procedures outlined in ANSI C63.5-1998. The chamber is 6.4 m (21 ft) wide × 9.8 m (32-ft) long × 6.1 m (20 ft) high and is lined with a ferrite/dielectric hybrid absorber. Other examples of this class of chamber are shown in Figure 10.9. This chamber uses grid ferrite tile throughout the installation.

Figure 10.8 Three-meter EMC chamber using hybrid ferrite tile absorber. (Photograph courtesy of Lehman Chambers, Inc., Chambersburg, PA.)

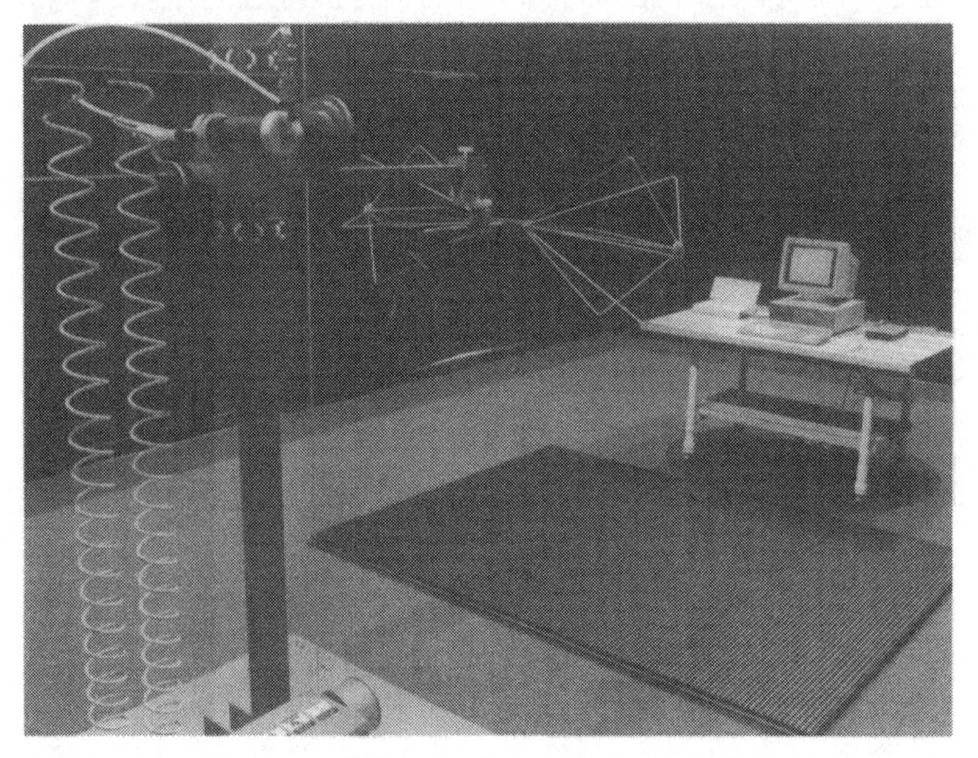

Figure 10.9 Three-meter EMC chamber using ferrite grid absorber. (Photograph courtesy of Panashield, Inc., Norwalk, CT.)

These chambers provide full compliance testing for ANSI C63.4-1992, EN 50147-2 and CISPR 22 requirements.

At 3-m range lengths, these chambers provide a typical normalized site attenuation (NSA) deviation of less than ±4 dB from theoretical NSA within the quiet zone, over the frequency range of 30 MHz to 1 GHz, when flat ferrite or grid tile are used. The quiet zone at the 3-m test distance is a cylinder up to 2.0 m in diameter, in accordance with the volumetric test procedure of ANSI C63.4-1992. Those chambers that use the ferrite/dielectric hybrid absorber permit testing beyond 18 GHz.

This class of chamber is also capable of performing full compliance testing for radiated immunity per IEC 61000-4-3, ENV 50140, and SAE J-1113 test requirements.

At 3-m range lengths, field uniformity of 0 to + 6 dB is achieved in the test aperture over the frequency range of 80 MHz to 2 GHz. The test aperture is a vertical plane 1.5 m by 1.5 m at an elevation of 0.8 m to 2.3 m above the ground plane, outlined in the procedures given in EN 61000-4-3. This procedure can be extended to 18 GHz. Field intensities up to 200 V/m are usual for this type of test facility.

Ten-Meter Chambers. There are several sources for 10-m EMC chambers. Recent examples are shown in Color Plate 8 and Figure 10.10. These chambers pro-

Figure 10.10 Ten-meter EMC chamber using hybrid ferrite tile absorber. [Photograph courtesy of EMC Test Systems (ETS), Austin, TX.]

vide full compliance testing capability at a 10-m range length in accordance with most international regulations. The overall footprint is on the order of 19 m × 13 m. The height is typically 9 m. A majority of the chambers utilize the unique properties of ferrite and dielectric absorbers, and they incorporate a variety of hybrid configurations to optimize the working space. Up to four receiving antenna masts can be installed for high-rate compliance testing.

Most enclosures use double steel laminated RF shielding systems (prefabricated). This type of predesigned system has been the most widely used shielding system in the world for over 20 years, with hundreds installed each year.

When making an evaluation of product EM susceptibility in accordance with EN 6000-4-3, a series of prefabricated ferrite absorber panels are placed over the specular floor area attenuating the floor reflections from the chamber.

When measuring product radiated emissions, the portable ferrite panels are removed from the chamber. The chamber will simulate an open-area test site, or OATS. At frequencies from 30 MHz to 18 GHz, the site attenuation of the chamber will be within ±4 dB of the normalized site attenuation (NSA) of an OATS, as required by ANSI C63.4, SAE J-551, EN 55022, and EN 50147-2. Chambers of this size will accommodate the testing of equipment under test (EUT) up to 4 m in diameter and 2 m in height.

An example of a production test facility is illustrated in Figure 10.11 where two turntables are available in a time-shared 10-m range. As testing is being con-

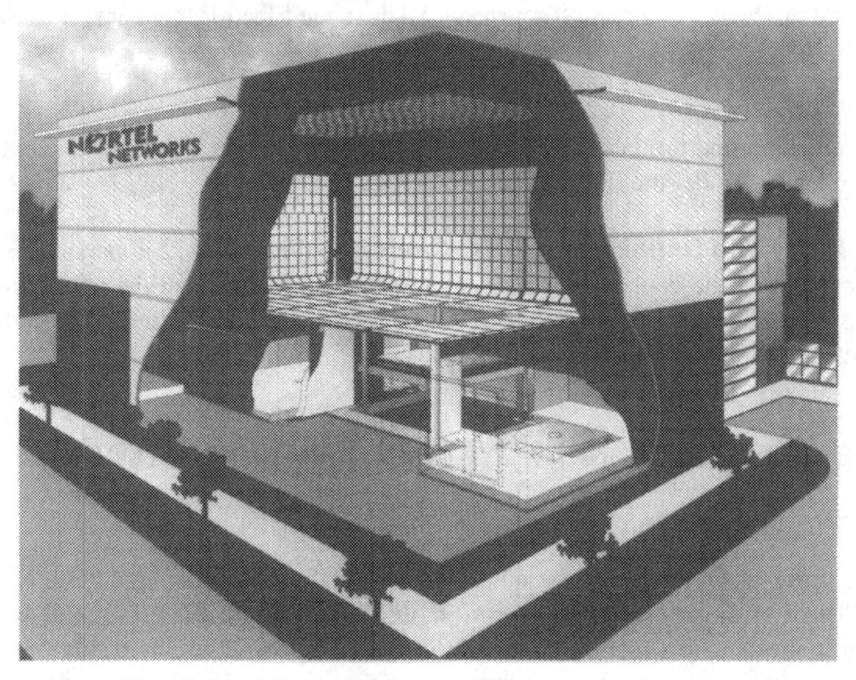

Figure 10.11 Concept drawing of a dual turntable 10-m EMC chamber. [Drawing courtesy of EMC Test Systems (ETS), Austin, TX.]

Figure 10.12 Interior view of Ten-meter Double Horn EMC test facility used to test automobiles. (Photograph courtesy of Advanced ElectroMagnetics, Inc., Santee, CA.)

ducted using one turntable, the other is prepared for the next test in the control room located under the floor of the 10-m range facility.

10.4.2.2 The Double Horn EMC Chamber. Figure 10.12 illustrates an interior view of the double horn EMC chamber. The geometry of the enclosure is shaped to reduce the surface area of the chamber surface and optimize the performance of the side-wall absorbing materials.

10.5 ELECTROMAGNETIC SYSTEM COMPATIBILITY TESTING

10.5.1 Introduction

Two different facilities will be described which illustrate the range of facilities dedicated to system testing. The first is dedicated to aircraft testing and is the largest anechoic test facility in the world. The second is more modest but represents a very important measurement requirement in our current world of satellite communications.

Color Plate 1. The absorber layout in a typical rectangular chamber is illustrated. (Photograph courtesy of Cuming Microwave Corporation, Avon, MA.)

Color Plate 2. The wedge absorber used in the tapered section is shown in a typical tapered chamber. (Photograph courtesy of Cuming Microwave Corporation, Avon, MA.)

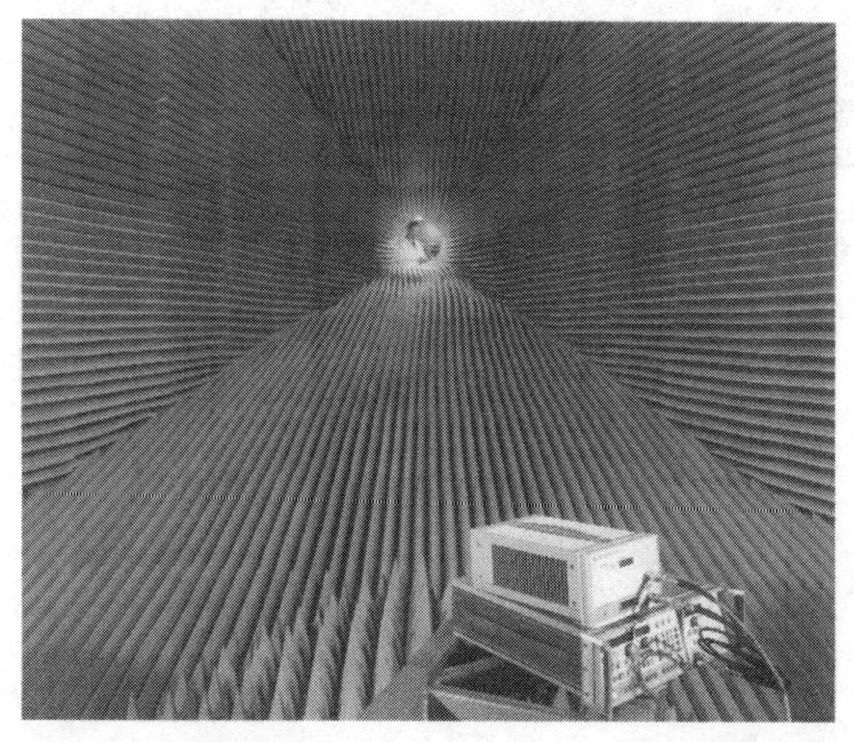

Color Plate 3. Same chamber as in Color Plate 2, looking up the taper.

Color Plate 4. Typical prime focus compact range chamber showing the arrangement of the absorbers on the chamber interior surfaces. (Courtesy of Lehman Chambers, Chambersburg, PA.)

Color Plate 5. Note the mirror-like performance of the reflectors in this dual reflector compact range system. (Courtesy of the Raytheon Company, El Segundo, CA.)

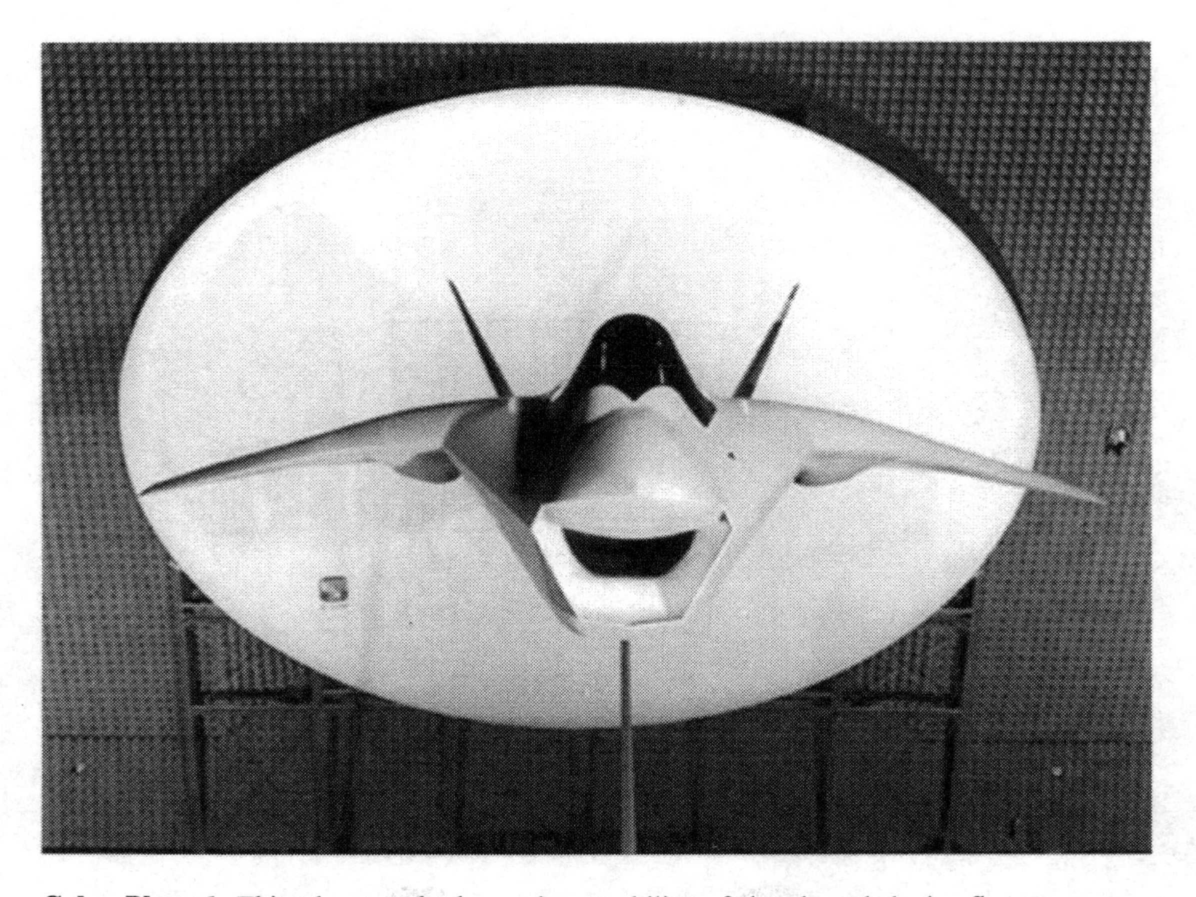

Color Plate 6. This photograph shows the capability of the shaped dual reflector compact range conducting RCS testing of full-scale fighter aircraft. (Courtesy of the Boeing Company, Seattle, WA.)

Color Plate 7. This large prime focus compact range is capable of RCS testing of 6-m-long targets. (Courtesy of MI Technologies, Norcross, GA.)

Color Plate 8. This 10-m EMC chamber uses Model FS-1500 hybrid ferrite absorber to achieve less than $\pm$ 3-dB site attenuation performance. [Courtesy of Electromagnetic Test Systems (ETS), Austin, TX.]

Color Plate 9. An interior view of a PIM anechoic test facility showing the arrangement for testing a large spacecraft antenna system. (Courtesy of the Harris Company, Melbourne, FL.)

10.5.2 Aircraft Systems

The U.S. Air Force maintains a very unique testing facility known as the Benefield Anechoic Facility (BAF) located at Edwards Air Force Base in California. This facility supports installed systems testing for avionics test programs requiring a large, shielded chamber with radio-frequency (RF) absorption capability that simulates free space. The BAF is an ideal ground test facility to investigate and evaluate anomalies associated with EW systems, avionics, tactical missiles, and their host platforms. Tactical-sized, single or multiple, and large vehicles can be operated in a controlled electromagnetic (EM) environment with emitters on and sensors stimulated while RF signals are recorded and analyzed. The largest platforms tested at the BAF have been the B-2 and C-17 aircraft. The BAF supports testing of other types of systems such as spacecraft, tanks, satellites, air defense systems, drones, and armored vehicles. See Figure 10.13 for a cutaway diagram of the facility. Another view of the chamber is shown in Figure 10.14 with various aircraft installed for test. The technical properties of the facility are given in Table 10.3.

The chamber is designed to support the following installed avionics systems/subsystems testing:

Antenna pattern radiation measurements (installed and stand-alone)

Electromagnetic interference/compatibility (EMI/C)

Figure 10.13 Cutaway view of the Benefield anechoic aircraft test facility. (Photograph courtesy of the U.S. Air Force, Edwards AFB, CA.)

Figure 10.14 B-1 and F-16 within the Benefield anechoic facility. (Photograph courtesy of the U.S. Air Force, Edwards AFB, CA.)

Table 10.3 Benefield Anechoic Chamber Parameters

Parameter	Data
Chamber size (L × W × H):	80.5 × 76.2 × 21.3 m (264 × 250 × 70 ft)
Range length:	28.0 m (92 ft)
Range type:	Free space
Equipment:	Large variety of high power sources and equipment for system simulation
Frequency coverage:	UHF, L, S, C, X, and Ku
Polarization coverage:	V, H, and circular
Angular coverage:	360 degrees
Turntable:	Diameter: 24.4 m (80 ft) Capacity: 250,000 lb Readout resolution: 0.05 degrees Turntable speeds: 0.1–0.6 degrees per second
Hoists:	Two 80,000-lb hoists are located above the turntable
Chamber lining:	0.46-m (18-in.) pyramidal absorber, ceiling, and walls, up to 1.22-m (48-in.) absorber is used on the turntable
Test region:	4.6–10.7 m (15–35 ft) above the turntable
Chamber anechoic properties:	Return loss as follows: 500 MHz, 72 dB; 1 GHz, 84 dB; 2 GHz, 96 dB; and 3–18 GHz, >100 dB

Source: Courtesy of U.S. Air Force, Edwards Air Force Base, CA.

Radiated/conducted emissions
Avionics and weapons/munitions/sensors integration
Systems sensitivity measurements
Controls and displays evaluation
Multiple aircraft interoperability

The chamber is also capable of performing electronic warfare (EW) system testing:

Free-space signal generation identification
Repeater and noise jamming
System response time
EW/avionics software development
Angle of arrival measurement
Free-space GPS jamming and spoofing
Minimum discernible signal
Target resolution/signal correlation
Radar cross section/imaging

10.5.2.1 Chamber Description. The shielded chamber of the Benefield installation is constructed of 11-gauge steel, which was continuously welded for an RF-tight shield. The enclosure is 80.5 m (264 ft) wide and 76.2 m (250 ft) deep and 21.3 m (70 ft) high at the peak sloping to 20.7 m (68 ft) at the walls. The slope is to prevent water collection on the ceiling due to accidental discharge of the main hanger fire protection system.

The shielded door for the chamber, 61 m (200 ft) wide and 20.1 m (66 ft) high, is a single panel door, which slides into place on railway tracks spaced 3.0 m (10 ft) apart. The door uses triple inflatable RF seals on the chamber side of the doorframe. A second outer hanger door is used to provide weather protection.

10.5.3 Spacecraft Test Facilities

An example of a spacecraft test facility is the large chamber located at the Harris Facility in Melbourne, Florida. The anechoic test facility has been successfully qualified for PIM (passive intermodulation) testing of large deployable mesh reflector antennas (up to 30 m). The chamber interior is shown in Color Plate 9. The chamber properties are given in Table 10.4.

Another spacecraft test facility is illustrated in Figure 10.15. This is a large tapered chamber used to test spacecraft for the Naval Research Laboratory in Washington, D.C.

Table 10.4 Properties of the PIM Spacecraft Antenna Test Facility

Parameter	Data
Operating frequency:	800 MHz to 60 GHz.
Working dimensions (W × L × H):	32.0 × 32.0 × 20.4 m, (105 × 105 × 67 ft).
Absorbers:	Uniform treatment of 24-in. pyramidal.
PIM:	Unique wall, ceiling, lighting, and air supply constructions provide high ambient RF isolation in a PIM free test environment. Measurements down to –155 dBm are possible.
Access:	Ceiling access with 4.3-m (14-ft) by 4.3-m (14-ft) opening and a 1.52-m (5-ft)-wide catwalk around the perimeter. Two 3-m (10-ft)-wide by 9-m (30-ft)-high swing doors and three personnel doors.
Hoist:	Five-ton hoist in center opening.
Clean room:	Meets class 100,000 clean room requirements.
Staging area:	Supported by an adjacent indoor staging area.
Control room:	Large adjacent control room with all supporting instrumentation.

Source: Courtesy of Harris Corporation, Government Systems Division, Melbourne, FL.

Figure 10.15 Large tapered chamber spacecraft test facility. (Photograph courtesy of Naval Research Laboratory, Washington, D.C.)

REFERENCES

1. E. F. Knott et al., *Radar Cross Section,* Artech House, Norwood, MA, pp. 392–394, 1985.
2. D. W. Hess, and K. Miller, Serrated-Edge Virtual Vertex Compact Range Reflectors, *Antenna Measurement Techniques Association Proceedings,* p. 6–19, 1988.
4. I. J. Gupta et al., Design of Blended Rolled Edges for Compact Range Reflectors, *Antenna Measurement Techniques Association Proceedings,* p. 59, 1987.

APPENDIX A

PROCEDURE FOR DETERMINING THE AREA OF SPECULAR ABSORBER TREATMENT

A.1 INTRODUCTION

The area to be covered with high-performance absorber in the specular region of an anechoic chamber is a function of the chamber geometry, the lowest operating frequency, and the required performance of the absorber.

A.2 FRESNEL ZONE ANALYSIS

For the purposes of establishing anechoic chamber design criteria, the reflection of electromagnetic waves, which illuminate typical chamber surfaces, are conveniently studied in terms of "zones of constant phase," or Fresnel zones, on the surface. For source and test heights, h_t and h_r, with a separation R between the bases of the antenna support structures, the shortest path between the source and test points via the chamber surface is

$$R_{RD} = [R^2 + (h_r + h_t)^2]^{1/2} \quad \text{(A-1)}$$

For vanishingly small wavelengths, this path would define the point of specular reflection (the center of the region of constant phase) at which the grazing angle ψ is given by

$$\varphi = \tan^{-1}[(h_r + h_t)/R] \quad \text{(A-2)}$$

If we establish a coordinate reference, as shown in Figure A.1, the path length via any other point (o, y, z) on the surface is written

$$R_R = [h_t^2 + y^2 + z^2]^{1/2} + [h_r^2 + y^2 + (R - z)^2]^{1/2} \tag{A-3}$$

Because $R_R > R_{RD}$, the phase of a ray traveling along R_R will lag that of the ray along R_{RD} by Δ_Φ radians, where

$$\Delta_\Phi = 2\pi/\lambda(R_R - R_D) \tag{A-4}$$

By definition, the locus of points (o, y_i, z_i) for which

$$\Delta_{\Phi i} = N\pi, \; N = 1, 2, 3, \ldots \tag{A-5}$$

or correspondingly

$$R_{Ri} - R_{RD} = N\lambda/2 \tag{A-6}$$

determines the outer boundary of the Nth Fresnel zone. The inner boundary of the Nth Fresnel zone is given by

$$R_{Ri} - R_{RD} = (N - 1)\lambda/2 \tag{A-7}$$

The definition are seen to be such that energy arriving at the test points from the outer bound of the Fresnel zone lags in phase by π radians with that energy arriving from the inner bound of the zone.

In anechoic chamber design problems, the pertinent parameters of a given Fresnel zone are the center, length, and width of the outer bound. These parameters can be calculated from equation (A-6). This is rewritten in terms of the chamber dimensions and coordinates as

$$[h_t^2 + y^2 + z^2]^{1/2} + [h_r^2 + y^2 + (R - z)^2]^{1/2} - [R^2 + (h_r + h_t)^2]^{1/2} = N\lambda/2 \tag{A-8}$$

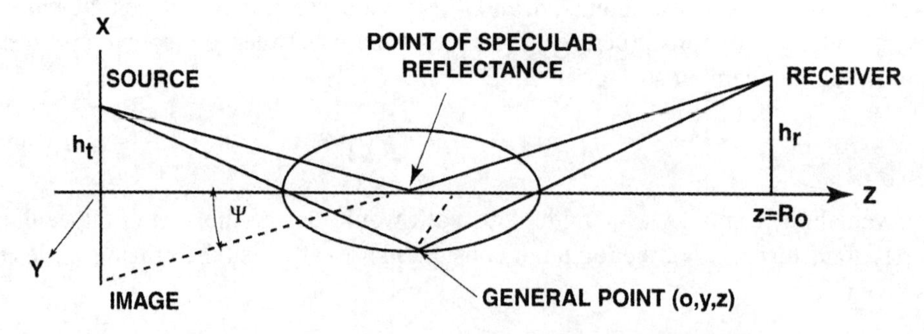

Figure A.1 Sketch for Fresnel zone boundary on a planar chamber surface.

This expression shows that the successive outer bounds of the Fresnel zones on a planar chamber surface describe a set of expanding ellipses whose major axis lies along the longitudinal chamber axis. The algebraic manipulations for solution of the desired parameters are simplified by definition of the following functions:

$$F_1 = (N\lambda/2R + \sec(\varphi)) \tag{A-9}$$

$$F_2 = (h_r^2 - h_t^2)/[(F_1^2 - 1)R^2] \tag{A-10}$$

$$F_3 = (h_r^2 + h_t^2)/[(F_1^2 - 1)R^2] \tag{A-11}$$

It can be shown that the following expressions result from the parameters of the outer bound of the Nth Fresnel zone, where the center is measured from the base of the transmit antenna location:

$$\text{Center:} \quad C_N = R(1 - F_2)/2 \tag{A-12}$$

$$\text{Length:} \quad L_N = RF_1(1 + F_2^2 - 2F_3)^{1/2} \tag{A-13}$$

$$\text{Width:} \quad W_N = R[(F_1^2 - 1)(1 + F_2^2 - 2F_3)]^{1/2} \tag{A-14}$$

Because practical anechoic chamber surfaces are not true planes, and the wavelengths at microwave frequencies do not satisfy the conditions of geometrical optics, the above expressions are not exact. However, they have been found to be useful in formulating practical anechoic chamber designs.

Design Example (Dimensions rounded off):
Assume a low-end operating frequency of 2 GHz.

Range length $R = 21$ ft.
Source height $h_t = 7$ ft.
Lowest receive $h_r = 5$ ft.

Assume $N = 6$. Experience has shown this is a minimum number of Fresnel zones that need to be covered to achieve good reflectivity in an anechoic chamber.

Then: $L = 15$ ft, $W = 8$ ft, and $C = 11$ ft. Then center of the specular region is located 11 ft from the transmitter wall. This places the start of the specular region right at the very edge of the test region.

APPENDIX B

TEST REGION AMPLITUDE TAPER

B.1 INTRODUCTION

The following analysis determines the amplitude taper in a rectangular chamber as a function of the source antenna beamwidth, width of the test region, and range length.

Let

$$R = \text{range length in feet}$$

$$W = \text{test region in feet}$$

In order to determine the antenna pattern, it is necessary to know its half-power beamwidth in both the E-plane and H-plane, generally specified by the antenna manufacturer.

Assume a factor n to conduct a trial calculation of the pattern function of the source antenna.

Let

$$a = 0, 1 \ldots, 4$$

$$n_a = 20, 22, 24, 26, 28$$

$$i = 0, 1, \ldots, 90$$

$$\varphi_i = 0 + i$$

$$PAT_{i,a} = 10\log[\cos(\varphi_i/57.3)^{n_a}]$$

Solving this equation yields the set of curves given in Figure B.1.

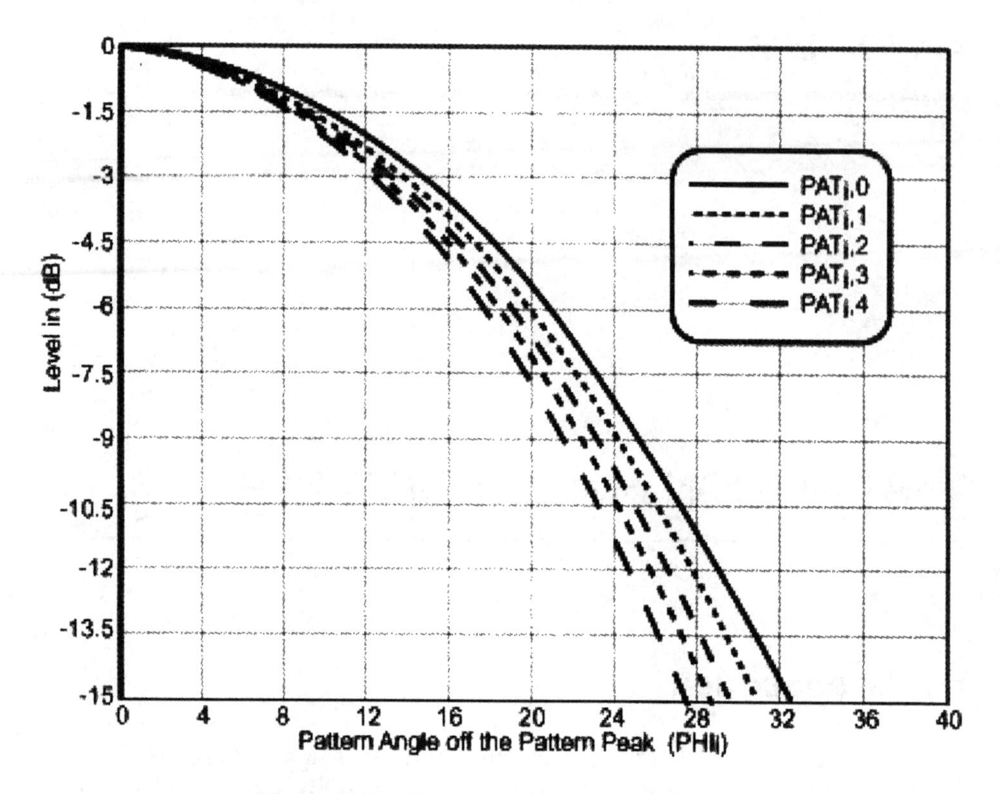

Figure B.1 Source antenna pattern levels.

From these data, pick an n factor from the family of curves that intersects the –3-dB line at the half-power point on the source antenna pattern. These same curves are used to determine the pattern level of energy illuminating the specular region of the anechoic chamber.

Assume $N1 = 21$ for a pattern factor in azimuth and $N2 = 21$ for a pattern factor in elevation.

Compute the angle to the edge of the test region.

$$ANG = W/(2R)$$

$$ANG = 0.125$$

$$TANG = a\tan(ANG)57.3$$

$$TANG = 7.126$$

$$TAPERAZ = 10\log[\cos(TANG/57.3)^{N1}]$$

$$TAPERAZ = -0.707 \text{ dB}$$

$$TAPEREL = 10\log[\cos(TANG/57.3)^{N2}]$$

$$TAPEREL = -0.707 \text{ dB}$$

B.2 ANTENNA DATA

As a matter of reference, the two most common standard horn manufacturers state that their products have the following properties.

The Narda company specifies their line of standard gain horns as follows:

All horns have a nominal gain of 16.5 dBi and a gain variation of 1.5 dB over the band. The HPBW run 34 to 23 degrees, low to high frequency. The nominal HPBW is 28 degrees. Sidelobes in the H-plane are all greater than 20 dB. Sidelobes in the E-plane run as follows: first lobe 13 dB, second 18 dB, and the remainder are all greater than 20 dB.

MI Technologies, formerly Scientific Atlanta, Inc., specify their line of standard gain horns as follows:

SGH 1.7 nominal gain: 15.5 dBi; nominal HPBW: EP 30 degrees, HP 27 degrees

SGH 2.6 nominal gain: 18 dBi; nominal HPBW: EP 23 degrees, HP 22 degrees

SGH 3.9 nominal gain: 18 dBi; nominal HPBW: EP 23 degrees, HP 22 degrees

SGH 5.8 nominal gain: 22.1 dBi; nominal HPBW: EP 12 degrees, HP 13 degrees

SGH 8.2 nominal gain: 22.1 dBi; nominal HPBW: EP 12 degrees, HP 13 degrees

SGH 12.4 nominal gain: 24.7 dBi; nominal HPBW: EP 9 degrees, HP 10 degrees

BAE Systems, formerly AEL, specify their ridged horn antennas as follows"

H4901 and H5001 nominal beamwidth: EP 30 degrees, HP 35 degrees

APPENDIX C

DESIGN/SPECIFICATION CHECKLISTS

C.1 INTRODUCTION

The following anechoic and enclosure design/specification checklists are provided to ensure that a systematic approach is made to the design or procurement of electromagnetic anechoic chambers. Because the enclosure design is a function of the overall chamber design this checklist is provided last in Section C.5. This appendix follows the format of the book in that the design of rectangular anechoic chambers is presented first, then the compact range, and finally chambers where the enclosure geometry is part of the electromagnetic design.

The following general checklist is given to help make sure that all factors in a given anechoic requirement are identified.

1. Determine what measurements are to be performed in the test facility.
   a. Antenna patterns
   b. Gain
   c. Cross-polarization
   d. Axial ratio
   e. Radomes
   f. Radar cross section
   g. System Isolation
   h. Electromagnetic emissions
   i. Electromagnetic immunity
   j. Impedance

   k. Hardware-in-the-loop
   l. Other
2. Determine type of chamber required.
   a. Far field
   b. Compact range
   c. Near-field range
   d. EMC test facility
   e. Impedance testing
   f. System integration
   g. PIM test facility
   h. Hardware-in-loop facility
   i. Other
3. Determine geometry of the required chamber (Refer to suggested chamber selection guide, Figure C.1)
4. Determine who is doing the installation, owner or contractor.
5. Determine if the chamber performance is to be verified.

C.2 THE RECTANGULAR CHAMBER

C.2.1 Introduction

The rectangular chamber can take many forms, depending on the type of measurement to be conducted in the chamber. The following is arranged by the type of test to be conducted.

C.2.2 Antenna Testing

The following is a list of items that should be included in a procurement specification for an anechoic chamber to be used for microwave antenna testing.

1. Use the enclosure checklist found in Section C.5 to specify the type of enclosure that is required to house the anechoic chamber. Also specify the type of fire protection system required. Generally, the facility insurance carrier sets the fire protection requirement.
2. Determine and specify the size of the antenna apertures to be tested and their frequency of operation. Specify the type of source antenna to be used in the chamber.
3. Determine the extraneous signal level that is required for the chamber, such as assuming –40 dB at the lowest operating frequency or by determining the maximum uncertainty that can be tolerated on a specific sidelobe level and operating frequency.

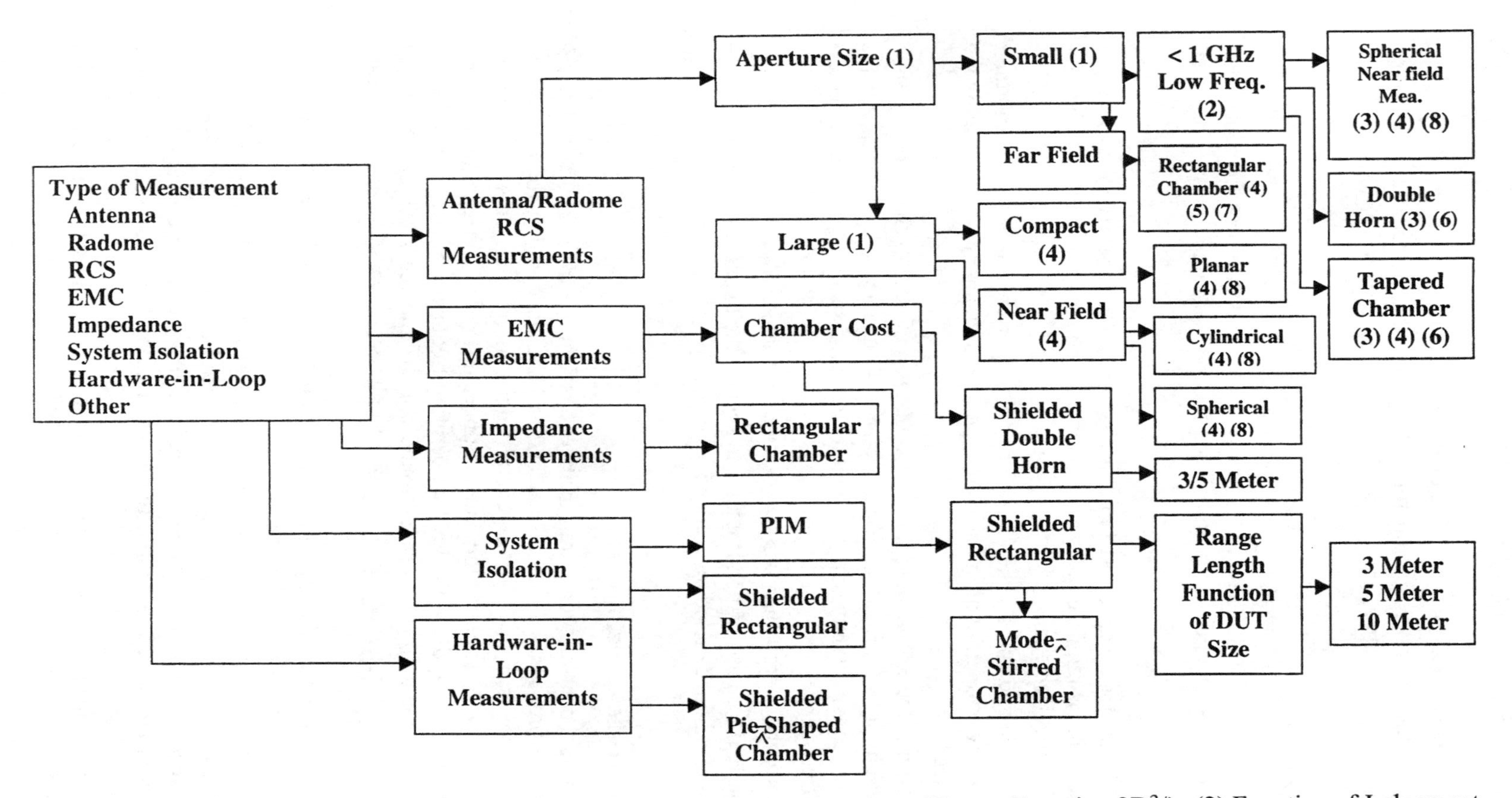

Figure C.1 Suggested chamber geometry selection guide. Notes: (1) Function of Range Equation $2D^2/\lambda$. (2) Function of Judgement and Budget. (3) Excellent Geometry for Wireless Measurements. (4) Shielded as required by application. (5) Can be used for wireless with careful design. (6) Can be used above 1 GHz with careful design. (7) Can be used below 1 GHz with careful design. (8) Generally, rectangular.

4. Specify the overall chamber size or space available.
5. Require the chamber manufacturer to specify the expected chamber size, chamber performance, and the detailed test procedure to verify it.

C.2.3 RCS Testing

The following is a list of items that should be included in a procurement specification for an anechoic chamber to be used for free-space microwave RCS testing.

1. Use the enclosure checklist found in Section 11.2 to specify the type of enclosure that is required to house the anechoic chamber. Specify the fire protection requirements.
2. Determine and specify the size of the RCS apertures to be tested and their frequency of operation. Specify the type of source antenna to be used in the chamber.
3. Determine the extraneous signal level that is required for the chamber, such as assuming –40 dB at the lowest operating frequency or by determining the maximum uncertainty that can be tolerated on a specific target and operating frequency.
4. Specify the overall chamber size or space available.
5. Require the chamber manufacturer to specify the expected chamber performance and the detailed test procedure to verify it.

C.2.4 Near-Field Testing

1. Near-field measurements are sensitive to reflections from the scanner structure, the probes used in the scanners, and the surrounding room in which the scanner is housed. As a general rule, it is recommended that the absorber used to terminate the antenna-under-test (AUT) have a minimum reflectivity of –40 dB at the lowest operating frequency of the near-field system. The remaining absorber around and behind the AUT can be lower performance material depending on the physical arrangement of the scanner in the measurement room.
2. The placement of the high-performance absorber is different for the three types of ncar-field test facilities. The planar range need only cover one wall behind the scanner with high-performance absorber to terminate the aperture under test. A lower-performance absorber may be used elsewhere. Whereas the cylindrical range requires the absorber to be placed on all wall surfaces around the antenna, the antenna scans these surfaces during the testing. The spherical near-field range requires a high-performance absorber on all chamber surfaces, because the main beam of the antenna scans all surfaces during the testing.

3. Careful modeling of the spherical near-field facility will provide insight as to the correct placement of absorber so as to control multipath reflections to acceptable levels.
4. The near-field range should be housed within a shielded enclosure if located where electromagnetic interference with other electronic equipment is possible.

C.2.5 EMI Testing

C.2.5.1 General Considerations

1. EMC chambers must be placed within shielded enclosures to isolate the measurements from the surrounding electromagnetic environment. A review of the checklists in Section C.5.5 will provide guidance on the best shielding solution for housing EMC chambers.
2. Determine the best path length for your measurement. The 3-m measurement distance is the most common, and most chamber manufacturers provide these types of test chambers. Specify the types of measurements that are to be conducted in the chambers and the size of your items to be tested. Be aware that the testing requirement for EMC is evolving at a somewhat fast rate, especially in Europe. Make sure that you have the latest requirements on hand prior to procuring a new test capability.

C.2.5.2 Emission Testing

1. For radiated emissions testing, the normalized site attenuation (NSA) deviation must be better than ±4 dB from theoretical NSA within the cylindrical test region over the frequency range 30 MHz to 1 GHz. The test procedure to be used is the volumetric test procedure of ANSI C63.4-1992.
2. The suggested room size is as follows for a 3-m emission test chamber:

Test Region Diameter	Length	Width	Height
2 m	6.4 m	6.1 m	6.1 m

3. Consider the following in sizing a 10-m chamber:

Test Region Diameter	Length	Width	Height
3 m (9.84 ft)	16.46 m	9.15 m	7.01 m
	54 ft	30 ft	23 ft
4 m (13.12 ft)	16.46 m	9.15 m	7.01 m
	54 ft	30 ft	23 ft
6 m (19.68 ft)	18.90 m	10.36 m	7.01 m
	62 ft	34 ft	23 ft

These are inside dimensions. If a raised floor is used, then add the height of the floor to obtain the shield-to-shield enclosure dimension.

C.2.5.3 Immunity Testing

1. Radiated immunity requires the field uniformity of 0, +6 dB to be achieved in the test aperture at a 3-m range length over the frequency range of 26 to 1000 MHz. The test aperture is a vertical plane 1.5 m × 1.5 m, at an elevation of 0.8 m to 2.3 m above the ground plane, and it follows the test procedure of EN 61000-4-3. If the chamber is dedicated to immunity testing, then the enclosure dimensions may be limited to a length of 6.13 m, the width of 5.79 m and the height of 5.4 m.

C.2.5.4 Mode-Stirred Test Facilities

1. Mode-stirred measurement techniques are being incorporated into the various government specifications. The size of the chamber determines the lowest operating frequency. The following versions are commercially available.

Test Volume	Frequency Range	Length	Width	Height
3.2 m^3	200 MHz to 18 GHz	4.83 m	3.61 m	3.05 m
44 m^3	80 MHz to 18 GHz	13.3 m	6.1 m	4.9 m

C.2.6 Isolation Testing

1. *PIM Chambers.* The PIM chamber is primarily used to test passive intermodulation products that can be generated by the structure of an antenna system. The effects of PIM have been observed in cell, personal communications systems (PCS), and spacecraft antenna installations. The chambers must be large enough to provide a combination of space loss and absorber reflectivity in excess of the PIM requirements. The interior surfaces of the chamber must be constructed such that passive intermodulation cannot be induced by the testing system.
2. *System Isolation.* System isolation chambers used for antenna-to-antenna coupling measurements on aircraft and spacecraft must be sized so that the sum of the space loss and absorber reflectivity exceeds the system isolation requirements.

C.2.7 Impedance Testing

1. Impedance chambers are used to measure the input impedance of antennas. The general guide is to design the chamber to have a reflectivity greater than the return loss (VSWR) of the antenna to be measured. To have a maximum error of less than 3 dB, the absorber reflectivity lining the enclosure should exceed the return loss by a minimum of 10 dB. The error charts given in Chapter 9 can be used to estimate the uncertainty for a given design.

C.3 COMPACT RANGE

C.3.1 Introduction

The compact range consists of a reflector system used to establish the testing environment within a rectangular chamber. The ranges are primarily used for large aperture radome, antenna, and RCS testing. They are especially useful for production or high-volume testing.

C.3.2 Antenna/Radome Testing

1. Compact ranges are highly specialized test facilities. Only a limited number of manufacturers exist for these types of test systems. Once a specific reflector system has been identified, the buyer should consult with the manufacturer and request their facility requirements to house their specific reflector system. If that is not possible due to procurement constraints, then the guidelines given in Chapter 8 should be used to establish the chamber requirements. These, plus the type of reflector system to be used, should be given to the anechoic chamber supplier so that he can ensure that the chamber is designed to properly house the specific compact range system.
2. Specify the compact range per the manufacturer's recommendations. These are expensive specially ordered systems. Care should be taken to ensure that all measurement requirements are clearly defined, so that the manufacturer of the reflector system can properly determine what parameters are critical to his design requirements. Specify that the chamber should not degrade the performance of the reflector system and that extraneous signal level in the test region should be at least 10 dB lower than the uncertainty specified for the reflector system. In most cases, the reflector supplier will provide a chamber specification suitable for his particular reflector system.

C.3.3 RCS Testing

1. When the compact range is to used for RCS testing, the same design considerations as used for antenna measurements are applied with the addition that the type of instrumentation radar to be used should be specified so that consideration can be given to gating. This generally means that the distance from the test region to the back wall may need to be longer than is used in CW testing.

C.4 SHAPED CHAMBERS

C.4.1 Introduction

Shaped chambers are those where the enclosure geometry contribute to the reducing the reflectivity in the test region of the chamber. The exact geometry is a function of the type of testing, frequency of operation, and cost.

C.4.2 Tapered Chamber

1. The tapered chamber is an established design concept primarily used for antenna testing below 1 GHz. Operations well up into the microwave frequencies is possible with care design and construction.
2. The tapered chamber is a free-space chamber, in that the range equation determines the size of the aperture that can be tested at a given frequency. It is typical to make the test region larger than the aperture to be tested should it be necessary to offset the antenna under test. The test region is typically one-third the cross section of the width/height of the test chamber.
3. The test region (quiet zone) size, the lowest and highest operating frequency, and the minimum performance at the low-end frequency determine the geometry of a tapered chamber. These should be selected carefully to ensure that the requirements are not too stringently selected, because the cost and size of the installation is directly proportional to these requirements.
4. For most antenna measurements, the chamber need only be metal-lined. Electromagnetic shielding is not normally required unless high-power measurements are to be conducted in the chamber or the facility is co-located with other sensitive electronic equipment. In most cases, where shielding is required, building the taper within a rectangular shielded enclosure is less expensive than forming tapered portions out of shielding panels.
5. When the tapered chamber is to used for RCS testing, the same design considerations as used for antenna measurements are applied with the addition that the type of instrumentation radar to be used should be specified so that consideration can be given to gating. This generally means that the distance from the test region to the back wall may need to be longer than is used in CW testing.

C.4.3 Double Horn Chamber

C.4.3.1 Antenna Testing

1. Specify the reflectivity in –dB versus frequency over the band of interest. Make sure that the specification is consistent with the absorber performance curves. Specify the geometry of the chamber in general terms. Require the vendor to determine the exact dimensions.
2. Set the required test region size and minimum range length that is required. Require that the Free-Space VSWR Method be used for chamber acceptance at a minimum of five frequencies and for both polarizations. Require the vendor to specify the types of antennas used in the testing.
3. Specify the chamber performance as you would a standard rectangular chamber, but require the vendor to consider the double horn geometry in his quote with cost comparisons to the standard rectangular chamber.

C.4.3.2 *EMI Testing*

1. EMC chambers must be shielded enclosures to isolate the measurements from the surrounding electromagnetic environment. A review of the checklists in Section C.5.5 will provide guidance on the best shielding solution for housing EMC chambers.
2. Determine the best path length for your measurement. The 3-m measurement distance is the most common, and most chamber manufacturers provide these types of test chambers. Specify the types of measurements that are to be conducted in the chambers and the size of your items to be tested. Be aware that the testing requirement for EMC is evolving at a somewhat fast rate, especially in Europe. Make sure that you have the latest requirements on hand prior to procuring a new test capabilities.
3. For radiated emissions testing, the normalized site attenuation (NSA) deviation must be better than ±4 dB from theoretical NSA within the cylindrical test region over the frequency range 30 MHz to 1 GHz. The test procedure to be used is the volumetric test procedure of ANSI C63.4-1992.
4. The suggested room size is as follows for a 3-m emission test chamber.

Test Region Diameter	Length	Width	Height
2 m	6.4 m	6.1 m	6.1 m

5. Consider the following in sizing a 10-m chamber:

Test Region Diameter	Length	Width	Height
3 m (9.84 ft)	16.46 m	9.15 m	7.01 m
	54 ft	30 ft	23 ft
4 m (13.12 ft)	16.46 m	9.15 m	7.01 m
	54 ft	30 ft	23 ft
6 m (19.68 ft)	18.90 m	10.36 m	7.01 m
	62 ft	34 ft	23 ft

These are inside dimensions. If a raised floor is used, then add the height of the floor to obtain the shield-to-shield enclosure dimension. The actual dimensions are a function of the layout required to achieve the double horn geometry as determined by the supplier.

C.4.4 Hardware-in-the-Loop Testing

1. *Hardware-in-Loop Testing.* A careful analysis of how extraneous signals affect the performance of a missile seeker system must be conducted prior to specifying the performance of the anechoic test chamber used for missile flight simulation. A good guide is that the chamber extraneous signal level should be at least 10 dB below that level of signal coming in on the side-

lobes of the seeker antenna. If the extraneous signal level is too high, it can cause the seeker to wander off target by some degree of error. The 10-dB margin will also ensure a good test environment for jammer testing.

C.5 SHIELDING DESIGN CHECKLIST

C.5.1 Introduction

Shielding is only required if the ambient electromagnetic environment would interfere with the measurements within the chamber or the fields developed in the chamber would cause interference with adjacent electronic equipment. In some cases the shielding is provided to prevent electromagnetic ease dropping. In any case the level of shielding effectiveness should be carefully determined so as not to overspecify the requirement.

This section provides a brief checklist as a guide for developing a specification for each of the three basic types of shielding used in anechoic chambers: modular, welded, and architectural. The final section reviews the special requirements that anechoic chambers require when using conventional construction. Part of the material in the checklists refers to normal architectural design considerations required in any facility. Some overlap exists between the examples; all four should be reviewed prior to finalizing a given specification. The design, manufacture, installation, and testing of electromagnetic shielding is a well-developed technology. For complete construction details and theory of electromagnetic shielding several references are available [1–3].

C.5.2 Checklist for Prefabricated Shielding

C.5.2.1 Introduction. For the purposes of this example, the galvanized panel modular enclosure [1] has been selected. The checklist is divided into four sections normally found in architectural specifications. They are architectural considerations, electrical considerations, mechanical considerations, and shielding considerations.

C.5.2.2 Architectural Checklist

1. *Floors.* The floor slabs under a prefabricated shielded enclosure should be recessed if a flush threshold is desired at a door opening. The depth is sized for the floor finishes. The slab should be smooth and level.

C.5.2.3 Electrical Checklist

1. *Filters.* All wires entering the enclosure should be filtered using filters matched for the service involved.
2. *Power Distribution.* It is recommended that one power penetration is made to distribute power within the enclosure and this circuit should be protected by circuit breakers.

3. *Chamber Lighting.* The lighting fixtures should be of the top-hat type—tha is, be mounted exterior to the enclosure. The lights should be located off the chamber axis, and the number should be limited to that essential for safe chamber operations.

C.5.2.4 Mechanical Checklist

1. *Penetrations.* All penetrations require special treatment to maintain the shielding integrity.
2. *Method of Construction.* The shielding should consist of prefabricated panels consisting of plywood or particle board laminated with galvanized sheet metal on both sides. The panels are precut and assembled using special clamping hardware.
3. *Fire Protection.* See Section C.5.6.

C.5.2.5 Shielding Checklist

1. *Shielding Specifications.* Select the appropriate specifications for the installation.
   a. IEEE 299-1991: IEEE Standard Method for Measuring the Effectiveness of Electromagnetic Shielding Enclosures, IEEE Press, 1991.
   b. NSA 65-6: Shielded Enclosure for Communications Equipment: General Specification
2. *Shielding Effectiveness.* The shielding effectiveness of prefabricated shielding is generally specified as follows:

   Magnetic Field: 56 dB at 14kHz, increasing to 100 dB at 200 kHz.

   Electric Field: 100 dB from 200 kHz to 50 MHz.

   Plane Wave: 100 dB from 50 MHz to 10 GHz commonly extended through 18 GHz.

3. *Work Included.* The shielding installer should be required to supply, install, and test the entire shielded enclosure. All testing should be completed prior to the installation of the RF absorber. The shielding installer should coordinate all work to ensure that the shielding integrity is maintained during the entire installation process.

C.5.3 Checklist for Welded Enclosures

C.5.3.1 Introduction

The welded enclosure [1] can attain the highest performance of all the methods of achieving shielding. The welded seam, if properly done, is the optimum method of achieving a seam in a shielded enclosure. Assuming that the welding is done properly, which only a good-quality assurance program can ensure, the performance of a welded enclosure is determined by the penetrations used to provide the various services. Each type of penetration must be designed for the overall performance requirement, and this is especially true of doors.

C.5.3.2 Architectural Checklist

1. *Shielded Volume.* The cost of field welding is very high; thus, the chamber designer must carefully consider the total volume to be included.
2. *Doors.* The mechanical design of an RF door is extremely important for maintaining the electromagnetic performance over the life of the facility. The mechanical strength of the hinging mechanisms and the resistance to warpage must be demonstrated with allowance for the weight of the RF absorber.

C.5.3.3 Electrical Checklist

1. *Electrical Conduit.* The conduit from the filter to the shield must be circumferentially welded at all joints and penetrations into the RF filter cabinet.

C.5.3.4 Mechanical Checklist

1. *Penetrations.* All penetrations must be designed to maintain the integrity of the shield, guidance on welded shield penetrations is given in Ref. 1.
2. *Ventilation Penetrations.* Honeycomb vent structures must be used for this service. The frame of the vent structure must be welded into the RF shield with continuous seam welds.
3. *Fire Protection.* See Section C.5.6.

C.5.3.5 Shielding Checklist

1. *Shield Requirements.* Define the shielding performance based upon a thorough evaluation of the facility needs, not just choosing a blanket specification. The performance requirements may be stated in a tabular or curve format, but it must be stated that the performance throughout the entire frequency spectrum must meet the minimum requirements. The recommended useful life of a welded shielded enclosure should be 30 years. The shielding effectiveness should be specified to hold for a minimum of 3 years using maintenance procedures supplied by the installation contractor.
2. *Shielding Effectiveness.* The following is typical of the specifications for a welded enclosure.

Magnetic Field:	40 dB at 1 kHz, raising to 120 dB at 10 kHz
Electric Field:	120 dB, 10 kHz to 50 MHz
Plane Wave:	120 dB, 50 MHz to 18 GHz.

C.5.4 Checklist for Architectural Shielding

C.5.4.1 Introduction. Architectural shielding (a variety of types are possible [1]), must be detailed very carefully because it is designed into the structure of the facility. It is especially important that every penetration be defined and detailed since the overall shielding integrity is directly related to the soundness of the penetrations.

C.5.4.2 Architectural Checklist

1. *Layout.* The floor plan of the space to be shielded should be laid out with consideration given to minimizing the number of penetrations and RF doors.

C.5.4.3 Electrical Checklist

1. *Filters.* All RF filters should be mounted in filter cabinets, and a single penetration should be passed through the shield and clamped on both sides of the single shield. The penetration should be solder sealed on the inside.

C.5.4.4 Mechanical Checklist

1. *Penetrations.* All penetrations through the shield must be mechanically supported, and a good solder seal must be made between the flange of the penetration and the shielding. It is recommended that penetrations be grouped using a common metal plate for penetrations.
2. *Fire Protection.* See Section C.5.6.

C.5.4.5 Shielding Checklist

1. *Scope of Work.* Ensure that all work necessary to accomplish the shield is clearly defined. This includes the shield, all supporting structures, finishes, mechanical items, electrical items, and testing.
2. *Shielding Specifications.* The performance specifications need to be carefully defined and specified and depend on the type of materials used in the construction of the enclosure. Examples are given in Ref. 1.

C.5.5 Conventional Construction

1. Conventional construction is often used in the construction of anechoic chambers. The most common form of construction is to frame out the chamber using conventional wood or steel studding. The interior surfaces of the chamber are lined with plywood. This surface is metal lined using lightweight galvanized steel sheet or aluminum foil. The sheets are lapped and then taped. The metal liner backs up the absorber and provides a controlled reflecting surface within the chamber. The absorber is then attached using contact adhesives specially selected to work with the fire retarded foam absorber. Ferrite chambers use specially developed installation procedures to allow for the weight of the ferrite tile.
2. Another method of construction for foam chambers is to use moisture proof drywall. The metal foil side is placed on the inside surface of the chamber.

The absorber lining should not exceed 0.6 m (2 ft) in height or the weight of the absorber will peel the lining from the drywall.

3. Use the guidance given in Section C.5.6 for fire protection guidance.

C.5.6 Fire Protection

1. Define the fire protection requirements as specified by the insurer of the building in which the chamber is to be located and by the local fire department. Three versions are generally used. (1) Sprinkler systems: Care should be taken to minimize the effect of reflections from the standpipes by locating them off the chamber axis in the ceiling of the chamber. Recently, some suppliers have been using phenolic pipe in lieu of the normal black pipe. (2) Gas discharge systems: A number of Halon substitutes have been developed which smother fires but do not damage the facility. CO_2 has been used in vehicle test chambers, but care must be taken in the design of the fire protection system so that personnel are not caught in the chamber when the gas is discharged. (3) Containment: Some facility managers have determined that it is best to isolate the facility behind two-hour fire-rated walls and not use any fire protection equipment except smoke detectors within the chamber. With the current grade of absorbers, fires have rarely occurred, but a large number of chambers have been ruined by accidental discharge of water by sprinkler systems. The absorbers act like large sponges and are very difficult to dry out if soaked by water. Some recent shielded installations have used sprinklers above the chamber to enhance the containment.

REFERENCES

1. L. H. Hemming, *Architectural Electromagnetic Shielding Handbook,* IEEE Press, New York, 1992.
2. R. Morrison, *Grounding and Shielding Techniques,* John Wiley & Sons, New York., 1998.
3. L. T. Genecco, *The Design of Shielded Enclosures,* Butterworth-Heinemann, Woburn, MA, 2000.

GLOSSARY

Certain technical terms are commonly used to describe most electromagnetic test facilities. The following list of terms, symbols, and notations, with their context in which they are used, is given. A good source for antenna terms is given in Ref. 3. Another source of electromagnetic measurement terms is given in Ref. 4. General electromagnetic terms are found in Ref. 5.

Anechoic (anti-echo) chamber: A test chamber designed to reduce unwanted reflected energy, so that free-space electromagnetic measurements can be simulated to an acceptable degree.

Antenna: That part of a transmitting or receiving system that is designed to radiate or to receive electromagnetic waves.

Aperture (of an antenna): A surface, near or on an antenna, upon which it is convenient to make estimations of field values for the purpose of computing fields at external points.

Back wall: The chamber wall immediately behind the test region in an anechoic chamber.

Backscatter: (1) The scattering of waves back toward the source. (2) Energy reflected or scattered in a direction opposite to that of the incident wave.

Bistatic angle: The angle between an incident wave and the reflected wave from a surface, where the angle of incidence equals the angle of reflection.

Bistatic reflectivity: The reflectivity when the reflected wave is in any specified direction other than back toward the transmit antenna.

Cassegrain antenna: A folded parabolic antenna that uses a subreflector to illuminate the primary reflector. Used on compact ranges to establish a far-field condition in a short-range length.

Cavity modes: The standing wave modes generated within a metal box when radio-frequency energy is inserted into the box.

Chamber reflectivity: See Reflectivity. Specifies the performance of an anechoic chamber in decibels versus frequency within the specific test region or quiet zone.

Compact range: The compact range works on the principle that the field in the radiating near-field of a large aperture is uniform in phase and amplitude, limited only by the diffraction from the edges of the aperture.

Conductive cellular structure (CCS) absorber: Absorber made into cell-like structures and painted with conductive materials.

Dielectric absorbers: Absorbers generally made of carbon-loaded foam.

Dielectric constant: (1) The property that determines the electrostatic energy stored per unit volume for unit potential gradient. (2) The real part of the complex permitivity.

Dipole antenna: Any one of a class of antennas producing a radiation pattern approximating that of an elementary electric dipole.

Directivity (of an antenna) (in a given direction): The ratio of the radiation intensity in a given direction from the antenna to the radiation intensity average over all directions.

Effective radiated power (ERP): In a given direction, the relative gain of a transmitting antenna with respect to the maximum directivity of a half-wave dipole and multiplied by the net power accepted by the antenna from the connected transmitter.

Electrical boresight: The tracking axis as determined by an electrical indication, such as the null direction of a conical-scanning or monopulse antenna system or the beam-maximum direction of a highly directive antenna.

Emission: The electromagnetic energy released by a device-under-test (DUT) in an electromagnetic test, primarily associated with electromagnetic compatibility testing.

E-plane, principal: For a linearly polarized antenna, the plane containing the electric field vector and the direction of maximum radiation.

Extraneous signal: The unwanted energy arriving from any angle within a test region. The levels are determined by systematically evaluating the test region using the Free-Space VSWR Measurement Method.

Ferrite absorbers: Ceramic tile loaded with magnetic oxides that absorb magnetic field energy.

Far field: (1) That region of the field of an antenna where the angular field distribution is essentially independent of the distance from the antenna. In this region (also called the free-space region), the field has predominantly plane-wave character—that is, locally uniform distributions of electric field strength and magnetic field strength in planes transverse to the direction of propagation. (2) Far-field testing implies that the field in the test region is uniform in phase and amplitude. The various range concepts approximate this in different ways.

Far-field pattern: Any radiation pattern obtained in the far field of an antenna

Far-field range: Far-field testing implies that the field in the test region is uniform in phase and amplitude. In an anechoic chamber, it means that the range length is sufficient to provide a test environment that meets the far-field criteria.

Free-space voltage standing wave ratio (VSWR): A test method for evaluating the quality of a test region. It involves moving an antenna probe through the test region and recording the field variation versus position. The level of extraneous energy is a function of the magnitude of the ripple recorded and the level of the observation point with respect to the peak of the probe's pattern.

Gain: The ratio of the radiation intensity, in a given direction, to the radiation intensity that would be obtained if the power accepted by the antenna were radiated isotropically.

Grazing incidence: When an electromagnetic wave arrives parallel to the surface of a chamber wall.

Gregorian reflector antenna: A paraboloidal reflector antenna with a concave subreflector, usually ellipsoidal in shape, located at a distance from the vertex of the main reflector which is greater than the prime focal length of the main reflector.

Ground plane; imaging plane: A conducting or reflecting plane functioning to image a radiating structure.

GTEM: Gigahertz transverse electromagnetic modal: A testing device for radiated emissions and immunity testing of small-sized equipment.

Half-power beamwidth (HPBW): In a radiation pattern cut containing the direction of the maximum of a lobe, the angle between the two directions in which the radiation intensity is one-half the maximum value.

Hybrid dielectric absorber: All foam absorbers made of laminated sheets on the base, topped with a pyramidal absorber.

H-plane, principal: For a linearly polarized antenna, the plane containing the magnetic field vector and the direction of maximum radiation.

Immunity (to a disturbance): (1) The ability of a device, equipment, or system to perform without degradation in the presence of an electromagnetic disturbance. (2) A term used to define the level at which an electronic device can sustain a given electric or magnetic field strength.

Magnetic absorber: Absorbing materials containing iron or other magnetic metal materials, which in the right combinations absorb magnetic field energy. Generally used below 1000 MHz.

Near-field pattern: Any radiation pattern obtained in the near-field of an antenna.

Near-field range: The aperture near the antenna under test is probed for phase and amplitude information, and then the far field is calculated using a transform from the near-field to the far-field.

Normal incidence: When an electromagnetic wave arrives normal to the plane of the device-under-test or to an absorbing surface.

Open-area test site (OATS): Sites used to measure the emissions from electronic devices. Defined in ANSI C63.4-1992.

Passive intermodulation (PIM): A form of signal distortion that occurs whenever signals at two or more frequencies conduct simultaneously in a passive device. A special problem for spacecraft designed for communications. It is now showing up in components designed for wire-less communication base stations.

Pyramidal absorber: A pyramidal-shaped foam structure impregnated with conductive materials used for absorbing microwave energy.

Quiet zone (QZ): The described volume within an anechoic chamber where electromagnetic waves reflected from the walls, floor, and ceiling are stated to below a certain specified minimum. The QZ may have a spherical, cylindrical, or rectangular shape depending on the chamber characteristics. See test region.

Radar cross section: (1) For a given scattering object, upon which a plane wave is incident, that portion of the scattering cross section corresponding to a specified polarization component of the scattered wave. (2) A measure of the reflective strength of a radar target; usually represented by the symbol (and measured in square meters. RCS is defined as 4(times the ratio of the power per unit solid angle scattered in a specified direction of the power unit area in a plane wave incident on the scatterer from a specified direction. More precisely, it is the limit of that ratio as the distance from the scatterer to the point were the scattered power is measured approaches infinity. *Note:* Three cases are distinguished:

1. Monostatic or backscatter RCS when the incident pertinent scattering directions are coincident but opposite in sense.
2. Forward-scatter RCS when the two directions and senses are the same.
3. Bistatic RCS when the two directions are different. If not identified, RCS is usually understood to refer to case 1). In all three cases, RCS of a specified target is a function of frequency, transmitting and receiving polarizations, and target aspect. For some applications (e.g., statistical detection analyses), it is described by its average value (or sometime its median value) and statistical characteristics over an appropriate range of one or more of those parameters.

Range length: The distance between the source antenna and the center of the test region.

Receiving wall: Another term for the back wall of an anechoic chamber.

Reflectivity: (1) For a radio-frequency (RF) absorber, the ratio of the plane wave reflected density (P_r) to the plane wave incident power density (P_i) at a reference pint in space. It is expressed in dB as $R = 10 \log_{10} (P_r/P_i)$. (2) A term commonly used to specify the performance of an anechoic chamber. The term implies that the level of extraneous energy in the test region is at a maximum level. See Extraneous energy.

Sidelobe level, relative: The maximum relative directivity of a sidelobe with respect to the maximum directivity of an antenna usually expressed in decibels.

Standing waves: The waves set up within a metal cavity when radio-frequency energy is inserted within a metal cavity, such as an anechoic chamber.

Susceptibility: A term used to indicate the level of energy that a device-under-test must pass or can sustain. See Immunity.

Tapered chamber: A form of chamber used for low-frequency electromagnetic measurements.

Test region: The location within an anechoic chamber where the device under test (DUT) is located, also known as the Quiet Zone.

Transmitting wall: The wall on which a source antenna is mounted within an anechoic chamber.

Wavelength (λ): The distance between two points of corresponding phase of two consecutive cycles in the direction of the wave normal. The wavelength, λ, is related to the magnitude of the phase velocity, v_p, and the frequency, f, by the equation:

$$\lambda = v_p/f$$

In free space the velocity of an electromagnetic wave is equal to the speed of light, that is, approximately 3×10^8 m/s.

Wedge absorber: A foam structure loaded with conductive material shaped in a series of wedges and used extensively in RCS and tapered chambers.

Wide angle: See bistatic angle.

REFERENCES

1. W. H. Emerson, Electromagnetic Wave Absorbers and Anechoic Chambers Through the Years, *IEEE Transactions on Antennas and Propagation,* Vol. AP-21, No. 4, July 1973.
2. B. F. Lawrence, Anechoic Chambers, Past and Present, *Conformity,* Vol. 6, No. 4, pp. 54–56, April 2000.
3. IEEE Std 145-1983 *IEEE Standard Definitions of Terms for Antennas,* IEEE Press, New York, 1983.
4. ANSI C63.14: 1998—American National Standard Dictionary for Technologies of Electromagnetic Compatibility (EMC), Electromagnetic Pulse (EMP), and Electrostatic Discharge (ESD).
5. IEEE 100, *The Authoritative Dictionary of IEEE Standard Terms,* 7th edition, IEEE Press, New York, 2000.

SELECTED BIBLIOGRAPHY

Anechoic Chamber Design

G. Antonini and A. Orlandi, Three Dimensional Model Based on Image Theory for Low Frequency Correction Factor Evaluation in Semi-Anechoic Chambers, *Proceedings of the IEEE EMC Symposium,* pp. 549–554, 1997.

H. Anzai et al., Analysis of Semi-Anechoic chamber Using Ray-Tracing Technique, *Proceedings of the IEEE Symposium on EMC,* pp. 143–145, 1996.

A. E. Baker et al., A Method of Reducing the Number of Ferrite Tiles in an Absorber Line Chamber, *EMC York 99, Conference Publication,* No. 464, pp. 59–64, 1999.

R. E. Bradbury, The World's Largest Anechoic Chamber, *Antenna Measurement Techniques Association Proceedings (AMTA*),* pp. 7-47–7-53, 1989.

B. Bornkessel and W. Wiesbeck, Numerical Analysis and Optimization of Anechoic Chambers for EMC Testing, *IEEE Transactions on Electromagnetic Compatibility,* Vol. 38, No. 3, pp. 499–506, August 1996.

S. A. Brumley, A Modeling Technique for Predicting Anechoic Chamber RCS Background Levels, *AMTA Proceedings,* pp. 116–1231, 1987.

S. Christopher et al., Anechoic Chamber Related Issues for Very Large Automated Planar Near-Field Range, *Proceedings of the IEEE EMC Symposium,* pp. 75–82, 1997.

J. Chenoweth and T. Speicher, *Cylindrical Near-Field Measurement of L-Band Antennas,* Near-Field Systems, Carson, CA 1996.

J. Friedel et al., Consideration in Upgrading to Bigger and Better Near-Field Chambers, *AMTA Proceedings,* pp. 141–146, 1995.

S. Galagan, Understanding Microwave Absorbing Materials and Anechoic Chambers, *Microwaves,* Part 1, December 1969; Part 2, January 1970; Part 3, April 1970.

*AMTA will be used for the remainder of the listings.

H. Garn, Radiated Emission Measurements in Completely Absorber Line Anechoic Chambers without Ground-Planes, *Proceedings of the IEEE Symposium on EMC,* pp. 390–393, 1989.

J. A. Graham, Jr., A Performance vs. Cost Model for Anechoic Chamber, *AMTA Proceedings,* pp. 11B-3–11-B-4, 1990.

J. Haala and W. Wiesbeck, Upgrade of Foam Equipped Semi-Anechoic Chambers to Fully Anechoic Chambers by the Use of Ferrite Tiles, *Proceeding of the IEEE EMC Symposium,* pp. 14–19, 1988.

C. L. Holloway and E. F. Kuester, Modeling Semi-anechoic Electromagnetic Measurement Chambers, *IEEE Transaction on Electromagnetic Compatibility,* Vol. 38, No. 1, pp. 79–84, February 1996.

R. G. Immell, Cost Effective, High Performance Anechoic Chamber Design, *AMTA Proceedings,* pp. 122–127, 1987.

M. K. Mansour and J. Jarem, Anechoic Chamber Design Using Ray Tracing and Theory of Images, *IEEE Proceeding of EMC, Southeasterncon,* pp. 689–695, 1990.

S. R. Mishra, and T. J. F. Pavlasek, Design of Absorber Lined Chambers for EMC Measurement using a Geometrical Optics Approach, *IEEE Transactions on Electromagnetic Compatibility,* Vol. EMC-26, No. 3, pp. 111–119, August 1984.

Y. Naito et al., Criteria for Absorber's Reflectivity Lined in Semi-Anechoic Chambers using Ray-Tracing Technique, *Proceedings of the IEEE Symposium on EMC,* pp. 140–142, 1996.

A. Orlandi, Multipath Effects in Semi-Anechoic Chambers at Low Frequencies; a Simplified Prediction Model, *IEEE Transactions on Electromagnetic Compatibility,* Vol. 38, No. 3, pp. 478–483.

D. A. Ryan et al., *A Two Dimensional Finite Difference Time Domain Analysis of the Quiet Zone Fields of an Anechoic Chamber,* Pennsylvania State University, January 1992, Report to NASA, Hampton, VA.

Anechoic Chamber Evaluation

B. Bornkessel and W. Wiesbeck, Influence of Geometrical Asymmetries on Anechoic Chamber Performance *AMTA Proceedings,* pp. 189–193, 1994.

D. N. Clouston, P. A. Langsford, and S. Evans, Measurement of Anechoic Chamber Reflections by Time-Domain Techniques, *IEE Proceedings,* Vol. 135, No. 2, pp. 93–97, April 1988.

R. R. DeLyser et al., Figure of Merit for Low Frequency Anechoic Chambers Based on Absorber Reflection Coefficients, *IEEE Transactions on Electromagnetic Compatibility,* Vol. 38, No. 4, p. 576, November 1996.

R. F. German, Comparison of Semi-Anechoic and Open Field Site Attenuation Measurements, *Proceedings of the IEEE Symposium on EMC,* 1982, pp. 260–265.

K. Haner, Anechoic Chamber Evaluation, *AMTA Proceedings,* pp. 206–211, 1994.

E. W. Hess, Field Probe Measurements and Stray Signal Evaluation of a Spherical Near-Field Range, *AMTA Proceedings,* 1982.

T.-H. Lee et al., Analysis of Anechoic Chamber Performance, *AMTA Proceedings,* pp. 200–205, 1994.

R. A. McConnell and C. Vitek, Calibration of Fully Anechoic Rooms and Correlation with OATS Measurements, *Proceedings of the IEEE on EMC,* pp. 134–139, 1996.

S. Shastry et al., Reflectivity Level of RF Shielded Anechoic Chamber, *Proceedings of the IEEE Symposium on Electromagnetic Compatibility,* 1995, pp. 578–583.

Absorber Design

S. A. Brumley, Evaluation of Anechoic Chamber Absorbers for Improved Chamber Designs and RCS Performance, *AMTA Proceedings,* pp. 116–121, 1987.

S. A. Brumley, *Evaluation of Microwave Anechoic Chamber Absorbing Materials,* Master's thesis, Arizona State University, May 1988, AD-A198 055.

C. T. Dewitt and W. D. Burnside, Electromagnetic Scattering by Pyramidal and Wedge Absorber, *IEEE Transactions on Antennas and Propagation,* Vol. 36, No. 7, pp. 971–984, July 1988.

T. Ellam, An Update on the Design and Synthesis of Compact Absorber for EMC Chamber Applications, *Proceedings of the IEEE Symposium on Electromagnetic Compatibility,* 1994, pp. 237–241.

J. R. J. Gau and W. D. Burnside, Transmission Line Approximation for Periodic Structures of Dielectric Bodies and its Application to Absorber Design, *IEEE Transactions on Antennas and Propagation,* Vol. 45, No. 8, August 1997.

R. J. Joseph, A. D. Tyson, and W. D. Burnside, Absorber Tip Diffraction Coefficient, *IEEE Transactions on Electromagnetic Compatibility,* Vol. 36, No. 4, pp. 372–379, November 1994.

Y. Naito et al., Ferrite Grid Electromagnetic Wave Absorber, *IEEE Symposium on EMC,* pp. 254–259, 1993.

K. Walther, Reflection Factor of Gradual-Transition Absorbers for Electromagnetic and Acoustic Waves, *IRE Transactions on Antennas and Propagation,* November, pp. 608–621, 1960.

Sun Weimin et al., Analysis of Singly and Doubly Periodic Absorbers by FDFD Method, *IEEE Transactions on Antennas and Propagation,* Vol. 44, No. 6, pp. 798–805, 1966.

C. Yang et al., Microwave Absorber Performance Analysis from PMM Calculations and RCS Measurements, *AMTA Proceedings,* pp. 3A-9–3A-14, 1991.

Absorber Evaluation

J. L. Barnes, Flammability Test Procedure for Anechoic Foams, *AMTA Proceedings,* pp. 3-21–3-26, 1990.

D. Kermis et al., Tests of the Fire Performance of Microwave Absorber, *AMTA Proceedings,* 1990.

I. Montiel, INTA's Free Space NRL Arch System and Calibration for Absorber Material Characterization, *AMTA Proceedings,* 1995.

S. Tofani et al., A Time-Domain Method for Characterizing the Reflection Coefficient of Absorbing Materials form 30 to 1000 MHz, *IEEE Transactions on Electromagnetic Compatibility,* Vol. 33, No. 3, pp. 234–240, August 1991.

Related Antenna Design and Measurements

M. J. Alexander et al., Advances in Measurement Methods and Reduction of Measurement Uncertainties Associated with Antenna Calibration, *IEE Proceedings—Scientific Measurement and Technology,* Vol. 141, No. 4, pp. 283–286, July 1944.

ANSI C63.5-1998, American National Standard for Electromagnetic Compatibility-Radiated Emission Measurements in EMI Control-Calibration of Antennas, November 11, 1998.

R. Bansal, The Far-Field: How Far is Far Enough?, *Applied Microwave & Wireless,* Nov. 1999, pp. 58–60, Vol. 11, No. 11.

A. Braun et al., Wide Band Horn (ridged horn) Calibration by the Three Antenna Method, *Conference on Precision Electromagnetic Measurements,* pp. 425–426, 1996.

W. D. Burnside et al., An Ultra Wideband Low RCS Antenna for Chamber Applications, *AMTA Proceedings,* pp. 103–108, 1995.

W. D. Burnside et al., An Ultra-Wide Bandwidth Tapered Chamber Feed, *AMTA Proceedings,* pp. 103–108, 1996.

F. Gisin, Using ANSI C63.5 Standard Site Method Antenna Factors for Verifying ANSI C63.4 Site Attenuation Requirements, *IEEE Symposium on EMC,* pp. 313–314, 1993.

R. C. Hansen, Measurement Distance Effects on Sum and Difference Patterns, *AMTA Proceedings,* pp. 1-9–1-14, 1992.

IEEE Std 149-1979 IEEE Standard Test Procedures for Antennas, August 8, 1980.

A. A. Smith, Jr., Standard-Site Method for Determining Antenna Factors, *IEEE Transactions on Electromagnetic Compatibility,* EMC-24, No. 3, pp. 316–322, August 1982.

Open-Area Test Sites (OATS)

A. A. Smith, Jr. et al., Calculation of Site Attenuation from Antenna Factors, *IEEE Transaction on Electromagnetic Compatibility,* EMC-24, No. 3, pp. 301–316, August 1982.

Compact Range Design

I. J. Gupta et al., A Method to Design Blended Rolled Edges for Compact Range Reflectors, *Transactions On Antenna and Propagation,* AP-38, pp. 853–861, June 1990.

D. W. Hess et al., Virtual Vertex Compact Range Reflectors, *AMTA Proceedings,* pp. 2-32–2-37, 1989.

M. S. A. Sanad and L. Shafai, Design Procedure for a Compact Range Using Dual Parabolic Cylindrical Reflectors, *AMTA Proceedings,* pp. 137–142, 1986.

INDEX

D

E

F

G

H

I

ABOUT THE AUTHOR

Leland H. Hemming is a member of the technical staff of the Signature Development and Applications Group, Phantom Works, a design and development division of the Boeing Company, located in Mesa, Arizona. The group specializes in the development of antennas for Boeing's new military aircraft and missiles. Mr. Hemming began his involvement with antenna work repairing submarine antennas for the U.S. Navy in 1955. He obtained his B.S. in Physics from San Diego State University in 1963 and his MBA from National University, San Diego, California, in 1977. He has attended numerous short courses in applied electromagnetics and has been an active member of the IEEE since 1963, obtaining the status of Senior Life member in 2001.

He began his engineering career with Chu Associates in 1961, supporting their work on naval antennas. After graduation, he worked for Andrew Corporation in Orland Park, Illinois until 1965. He returned to California and worked for Bunker-Ramo in Canoga Park until 1967. At this time he joined Scientific Atlanta, Inc., in Atlanta, Georgia. At Scientific Atlanta, he worked on a variety of antenna and antenna instrumentation projects. This included working with anechoic chambers. In 1974, he again returned to California and joined Plessey Microwave Materials in San Diego as Chief Engineer and was responsible for their line of microwave materials and anechoic chambers. In 1981, Plessey chose to terminate the product line, so Mr. Hemming co-founded Advanced Electromagnetics, Inc., in San Diego, to continue the manufacture of microwave material and anechoic chambers. As Technical Director, Mr. Hemming designed the company's line of anechoic chambers. In 1983, he returned to corporate life and joined Global Analytics, Inc., of San Diego, California, as an antenna engineer. Global Analytics was acquired by Alcoa, Inc., which, in turn, sold the unit to McDonnell

Douglas. The group was located in Rancho Bernardo, a northern suburb of San Diego. In 1996, McDonnell Douglas terminated the San Diego business unit and moved the technical staff to its plant in Mesa, Arizona. In August 1999, Boeing acquired McDonnell Douglas and the unit was placed in the Phantom Works organization.

Mr. Hemming has been granted 10 patents in the field of applied electromagnetics, including antennas, radomes, electromagnetic shielding, and anechoic chambers. He has published numerous papers in these technical fields. His book *Architectural Electromagnetic Shielding Handbook,* published in 1992 by the IEEE Press, now in its second printing, is extensively used as a training manual in the shielding industry.

CPSIA information can be obtained
at www.ICGtesting.com
Printed in the USA
BVHW011455070719
552648BV00031B/188/P